Introduction to Nuclear Radiation Detectors

LABORATORY INSTRUMENTATION
AND TECHNIQUES

Series Editor: **Galen W. Ewing**

Seton Hall University

Volume 1: *The Laboratory Recorder*
By Galen W. Ewing and Harry A. Ashworth • 1974

Volume 2: *Introduction to Nuclear Radiation Detectors*
By P. J. Ouseph • 1975

A Continuation Order Plan is available for this series. A continuation order will bring delivery of each new volume immediately upon publication. Volumes are billed only upon actual shipment. For further information please contact the publisher.

Introduction to Nuclear Radiation Detectors

P. J. Ouseph

Department of Physics
College of Arts and Sciences
University of Louisville
Louisville, Kentucky

PLENUM PRESS • NEW YORK AND LONDON

Library of Congress Cataloging in Publication Data

Ouseph, P J 1933-
   Introduction to nuclear radiation detectors.

   (Laboratory instrumentation and techniques ; v. 2)
   Bibliography: p.
   Includes index.
   1. Nuclear counters. I. Title.
QC787.C6088   539.7'7   75-15744

ISBN-13: 978-1-4684-0837-9   e-ISBN-13: 978-1-4684-0835-5

DOI: 10.1007/ 978-1-4684-0835-5

©1975 Plenum Press, New York
Softcover reprint of the hardcover 1st edition 1975
A Division of Plenum Publishing Corporation
227 West 17th Street New York N.Y. 10011

United Kingdom edition published by Plenum Press, London
A Division of Plenum Publishing Company, Ltd.
Davis House (4th Floor), 8 Scrubs Lane, Harlesden, London NW10 6SE, England

To
Ann, Rosemary,
John, and Maryann

Preface

There have been many interesting developments in the field of nuclear radiation detectors, especially in those using semiconducting materials. The purpose of this book is to present a survey of the developments in semiconductor detectors along with discussions about gas counters and scintillation counters. These discussions are directed to detector users, usually scientists and technicians in different fields such as chemistry, geology, biochemistry, and medicine. The operation of these detectors is discussed in terms of basic properties, such as efficiency, energy resolution, and resolving time, which are defined in the first chapter. Differences among these detectors in terms of these properties are pointed out. Chapter 2, on interaction of radiations with matter, discusses how different radiations lose energies in matter and how differences in their behavior in matter affect the design and operation of detectors. Although emphasis is placed on fundamentals throughout the book, the reader is also made aware quite often of the new developments in the field of radiation detection.

The author has taught a course in radioisotopes for several years for science, engineering, medical, and dental students. The emphasis on topics varied from time to time to satisfy the varying interests of the students. However, the contents of this book formed the core of the course. About ten selected experiments on detectors were done along with this course (a list of these experiments may be supplied on request). The enthusiastic response received from the students encouraged the writing of this book.

The help I received from several individuals during the preparation of this book deserves special mention. Professor Manual Schwartz initiated and developed the course on radioisotopes and radiation detection at our university. As mentioned earlier, this book originated with this course, modified through years of teaching. Without his help in reading, correcting and modifying the manuscript, especially in its early stages, this book would not have been possible. I am deeply indebted to Professor Schwartz for this help and for his collaboration in writing two articles on detectors which gave me additional impetus in writing this book.

I should also acknowledge the help I received from Thomas Lanigan and Ralph Cutler of the editorial staff of Plenum Press. I am especially grateful to Dr. Galen Ewing, the editor of this series, for his patient efforts in correcting and editing this book. In preparing the diagrams, the help I received from Mrs. Catherine M. Bauscher deserves special recognition. Nor should I forget the assistance provided by Miss Lisa Hark in the typing of this manuscript. Finally, my appreciation goes to my family for their patience and encouragement.

Louisville, Kentucky   P. J. OUSEPH
May, 1975

Contents

1

Introduction

The history of nuclear radiation detectors parallels the growth in our knowledge of atomic and subatomic physics. An example is Roentgen's discovery of x rays in 1895 using photographic emulsions. Since that time emulsions have been widely used as radiation detectors. In the following year Becquerel discovered radioactivity with the help of photographic plates. Rutherford used a fluorescent screen and a telescope to view the flashes of light produced by alpha particles on the screen in his famous alpha-scattering experiments. His discovery of the nucleus was based on the results of these experiments. Similar detectors, known as "spinthariscopes," developed by Sir William Crookes in 1903 were the forerunners of the modern scintillation detectors.

It was also recognized that radiations produce ionization in air. Electroscopes and electrometers were used to detect the ionization produced by radiations, and these instruments were used in the discovery of cosmic rays. Marie Curie used electrometers in her study of the radioactivity of a wide variety of substances. Using these simple instruments, she was led to the conclusion that the activity was proportional to the quantity of the radioactive substance.

The extreme difficulty of counting fluorescent flashes prompted Rutherford and Geiger to look for a better detector. The facts that (1) radiations produce ions and (2) ions multiply in a sufficiently high electric field were used by them in the design

Table 1.1. Nuclear Radiation Detectors

Name	Type	Primary interaction	Medium
Ionization counter Proportional counter Geiger counter	Signal	Ionization	Gas
Scintillation counter	Signal	Excitation of electronic levels and associated production of photons	Gas, liquid, and solid
Semiconductor counters	Signal	Production of electrons and holes	Solid
Cerenkov counter	Signal	Production of photons by Cerenkov effect	Gas, liquid, and solid
Photographic emulsion	Track	Ionization	Solid
Cloud chamber (diffusion and expansion)	Track	Ionization	Gas
Bubble chamber	Track	Ionization	Liquid
Spark chamber	Track	Ionization	Gas and solid
Dielectric particle detector	Track	Ionization	Solid, tracks developed by etching
Photochromic detectors (2)	Track	Change in oxidation of Fe ions	Solid

of their new detectors. In 1908 they announced the perfection of a workable nuclear radiation detector. Since that time, there have been several important advances in methods of detection of nuclear radiation.

Table 1.1 lists the different types of detectors currently in use. The table underscores the truth of the statement made by McKay in 1953: "Whenever a nuclear physicist observes a new effect caused by an atomic particle, he tries to make a counter out of it."[1] The "track-type" counters are mainly of interest to high-energy-particle physicists. The "signal-type" counters are of interest to several categories of scientist, such as radiochemists, low-energy nuclear physicists, biologists, biochemists, and geologists using radioisotope techniques. Such counters are discussed in detail in Chapters 3–5.

The operation of a typical pulse counter is indicated in Fig. 1.1. A radiation (charged particle) incident on the counter produces a signal (output pulse). The proportionality between the amplitude of the signal (pulse height) and the energy of the radiation, as well as the efficiency of the counter, the energy resolution, and the time width of the pulse, are some of the factors one considers in selecting a counter. The characteristics that influence the selection of a counter are discussed below.

If the amplitude of the signal is directly proportional to the energy expended by the radiation in the detector, determination of the amplitude of the signal will enable one to measure the energy of the radiation.

The efficiency can be defined as the ratio of the number of detectable signals produced to the number of incident radiations. Sometimes, especially for low-energy radiations, the amplitude of the signals produced may be smaller than the noise, making them impossible to detect. Another factor is that the electronics used for analyzing the signals may have sensitivity limits, i.e., pulses below the lower limit will not be detected. Therefore the efficiency of a counter will vary with the lower limits of the electronics following the counter.

The energy resolution of a counter is the measure of its ability to distinguish between radiations of very close energies. This will depend on the spread in pulse heights for radiations of the same energy. A typical pulse height distribution is shown in

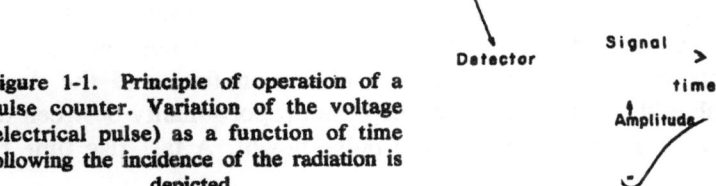

Figure 1-1. **Principle of operation of a pulse counter. Variation of the voltage (electrical pulse) as a function of time following the incidence of the radiation is depicted.**

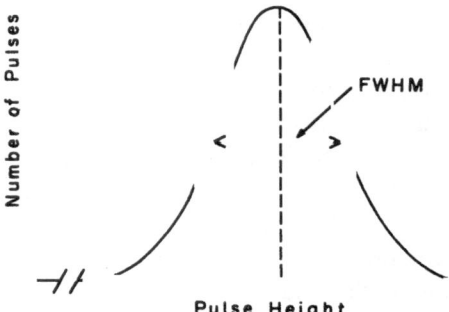

Figure 1-2.   Typical pulse height distribution (number
of output signals as a function of pulse height) for
radiations with the same energy. The position of the
peak for most detectors is proportional to the energy
and the full width of the distribution at half-maximum
depends on the type of detector.

Fig. 1.2. The resolution R, in percent, is defined as follows:

$$R = \frac{\text{full width at half-maximum}}{\text{pulse height at the peak}} \times 100 \qquad (1.1)$$

The width at half-maximum is the width of the pulse distribution
at half-height of the peak, as indicated in Fig. 1.2. The ability of a
counter to resolve between gamma rays increases as R decreases.

The time width of the pulse is another important factor
usually considered in selecting a counter. If the total width is
large, overlapping of pulses will take place, with a concomitant
loss of count. The pulse has two parts: (1) the rising part and (2)
the decaying part. The first part is determined by the properties
of the counter (such as the ion collection time) and the second
part depends on the time constant RC of the system. The total
width can be reduced by reducing the time constant; however, a
suitable relationship between the collection time and the decay
time should be maintained to keep the proportionality between the
energy of the radiation and the pulse height. A fast rise time for

the pulse is required for coincidence experiments with good time resolution.

Prior to a consideration of detectors, properties of some of the commonly occurring radiations will be described. This description will be followed by radioactive decay calculations and a discussion of interactions of radiation with matter. This material will serve as background to a consideration of counters in the remainder of the book.

REFERENCES

1. M. G. McKay, *Phys. Today* **6**, 10 (1953).
2. P. J. Ouseph, *Phys. Rev. Lett.* **30**, 1162 (1973).

2

Nuclear Radiations and Their Interaction with Matter

Detection of radiation is possible because in the interaction of radiation with matter, photons, electron–ion pairs, or electron–hole pairs can be produced. Gas counters (ionization chambers, proportional counters, and Geiger counters), cloud chambers, bubble chambers, and spark counters work because the radiations produce ions. In scintillation counters, radiation-induced excitation produces light quanta (photons) and in semiconducting counters, radiation produces electron–hole pairs. The number of ions, photons, or electron–hole pairs depends on the fraction of the energy of the radiation expended in the sensitive volume, on the properties of the material, and sometimes on the nature of the radiation. In the design of counters and in the analysis of information obtained from them, it is important to know how fast the radiations expend their energy in a medium, how much of it goes into producing ions, photons, or electron–hole pairs, what is the relationship between the number of ions and the energy, and how much material is needed to stop a radiation of a given energy. Some of these problems are discussed in this chapter. Neutral radiations like gamma rays and neutrons do not directly produce these effects and therefore their detection depends on an intermediate interaction in which charged particles are produced. Interactions of neutral radiations are discussed in the latter part of this chapter. However, it is appropriate at this point to become familiar with nuclear radioactivity, nuclear reactions, and terms such as decay constant, half-life, and reaction cross section. These topics are now discussed.

2.1. RADIOACTIVITY

Some nuclei undergo spontaneous decay or disintegration. This process is known as radioactivity. In radioactive decay, the energy per nucleon of the product daughter nucleus is less than that of the parent nucleus. A good fraction of the energy difference is usually carried away by radiations emitted during the decay. Based on the type of radiation released, there are three main classifications of radioactivity: (1) Alpha decay, where an alpha particle is released. An alpha particle consists of two protons and two neutrons and hence it is equivalent to the helium nucleus. (2) Beta decay, where the nuclear transformation involves emission of an electron (electron decay) or positron (positron decay) and sometimes capture of an electron (electron capture). In these three different types of beta decay, particles known as neutrinos or antineutrinos can be released. (3) Fission, splitting of the nucleus into two lighter nuclei, with a release of two or more neutrons and approximately 200 MeV of energy.

It is important to note a major difference between the energy distribution of alpha particles and that of beta particles. As pointed out earlier, a good fraction of the disintegration energy is carried away by the emitted radiations. The rest of the energy appears as the kinetic energy of the daughter nucleus, which recoils in such a way as to conserve momentum in the overall

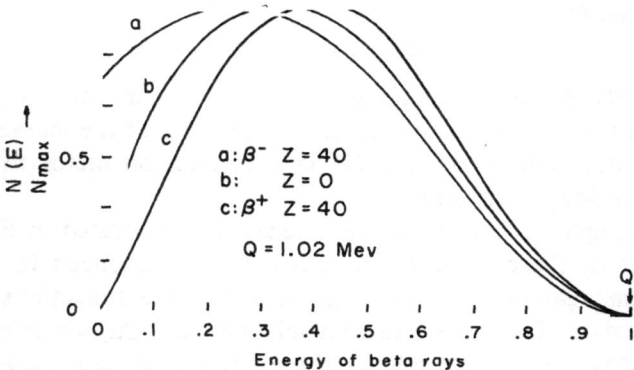

Figure 2-1. Energy distribution of beta rays. E_{max} is taken as 1.02 MeV which is equal to Q. Note the distortion at the low-energy end of the spectrum caused by electrostatic interaction between the beta rays and the nucleus. Curve *a* is for electrons, where the attraction between the electron and the nucleus reduces the energy of the particle; curve *b* assumes no interaction; curve *c* is for positrons, where repulsion increases the energy of most of the positrons.

process. In alpha decay there are only two reaction products, the daughter nucleus and the alpha particle, and therefore the disintegration energy can be distributed in only one way. Therefore, for a specific nuclear transition leading to alpha-particle emission, the energy of the alpha particle is always the same, given by the following equation:

$$E_\alpha = \frac{A - 4}{A} Q \qquad (2.1)$$

where E_α is the energy of the α particle, A is the atomic number of the parent nucleus, and Q is the disintegration energy.

In electron decay and positron decay the energy can be distributed among three particles, the nucleus, the electron or positron, and the antineutrino or neutrino, in an infinite number of ways. satisfying the principle of conservation of momentum. Therefore beta particles have a continuous energy distribution from zero to a maximum of Q, where Q is the disintegration energy (Fig. 2.1). The disintegration energy Q in MeV can be calculated from

$$Q = [M_p - (M_d + \Sigma m_r)] \times 931 \qquad (2.2)$$

where M_p is the mass of the parent nucleus in amu, M_d is the mass of the daughter nucleus, Σm_r is the sum of the masses of all the emitted radiations, and 931 is a conversion factor to obtain energy in MeV from amu.

Examples of alpha and beta decay are illustrated in Fig. 2.2. In most of these decays, the products are produced in excited states and gamma rays are emitted during the transitions to the ground state. There are a few nuclei (such as ^{64}Cu) which undergo radioactive decay through all three modes of beta decay. The 0.66-MeV gamma rays from ^{137}Cs and the 1.178- and 1.33-MeV gamma rays from ^{60}Co are used for calibration of gamma-ray spectrometers.

In electron capture there are only two products (the neutrino and the daughter nucleus) and therefore the energy distribution for the neutrino will not be continuous. However, the binding energy of the electron can vary, depending on the atomic shell in which the electron is located. For each binding energy, the neutrino will be emitted with a different energy, and consequently the neutrino can possess a discrete energy spectrum. The capture of an electron produces a vacancy in the atomic shell, and when this vacancy is filled an x ray characteristic of the product atom will be released. Detection of the characteristic x ray of the product atom is one way of observing the electron capture.

2.2. *RADIOACTIVITY CALCULATIONS*

Radioactive decay is a statistical one-shot process. The probability of one nucleus in a group of nuclei decaying in a finite time interval is independent of the time; in other words, the probability of decay at two different times t_1 and t_2 will be the same. Also, as far as an individual nucleus is concerned, it is impossible to predict when it will decay. However, one can calculate how many of a group of nuclei will decay in a time increment Δt at a given time.

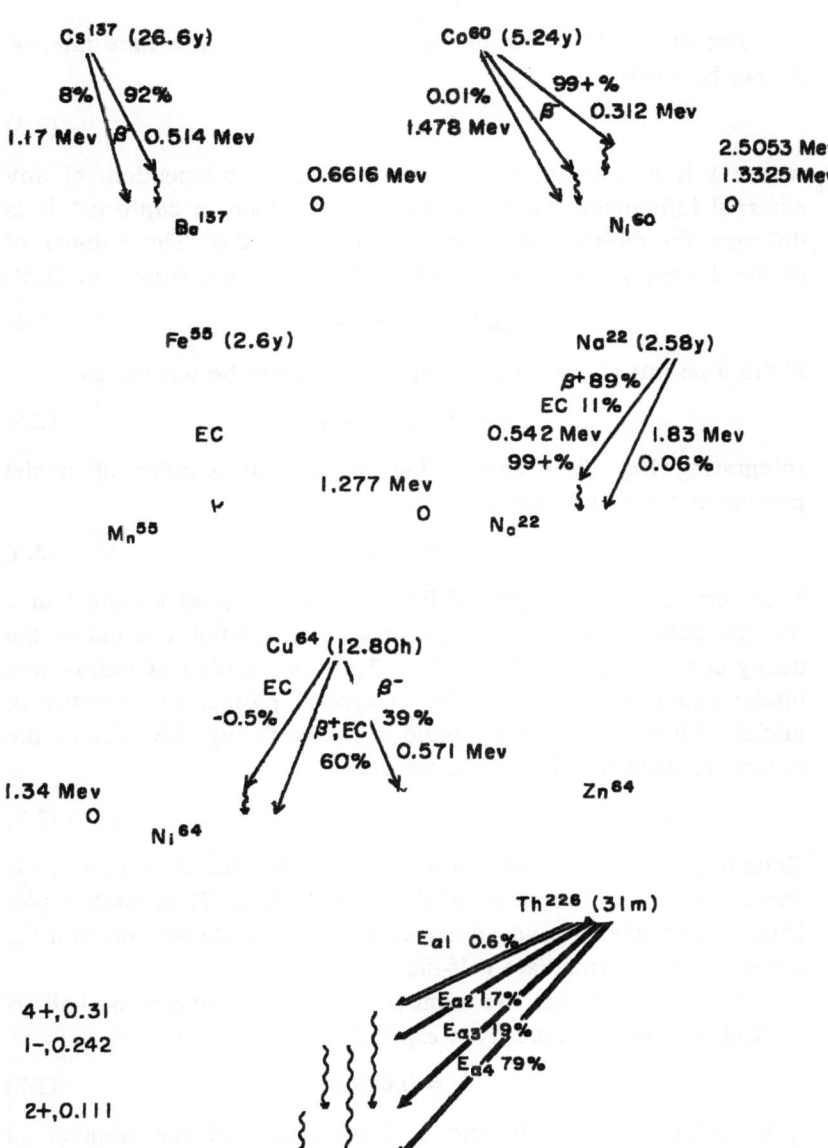

Figure 2-2. Decay schemes of ^{137}Cs, ^{60}Co, ^{55}Fe, ^{22}Na, ^{64}Cu, and ^{228}Th. Four groups of alpha particles (6.03, 6.1, 6.22, and 6.3 MeV) are released in ^{228}Th decay. β^-, β^+, and EC stand for electron decay, positron decay, and decay by electron capture respectively. A wavy arrow indicates gamma rays.

The probability P of decay of one nucleus in a time interval Δt can be written

$$P = \lambda \, \Delta t \tag{2.3}$$

where λ is a constant for an isotope and is independent of any external influence, except in the case of electron capture.* It is different for different nuclides and decay modes. The number of nuclei decaying from a group of N nuclei will be, from Eq. (2.3),

$$-\Delta N = NP = N\lambda \, \Delta t \tag{2.4}$$

If the time interval is very small, Eq. (2.4) can be written as

$$- \, dN/dt = \lambda N \tag{2.5}$$

Integrating Eq. (2.5) and taking N_0 as the number of nuclei present at $t = 0$, we obtain

$$N = N_0 e^{-\lambda t} \tag{2.6}$$

From dimensional analysis of Eq. (2.5) or Eq. (2.6) we see that λ has the dimension of reciprocal time. The symbol λ is called the decay constant. According to Eq. (2.6), the number of radioactive nuclei decreases exponentially. A typical plot of the number of nuclei with time on a linear scale is shown in Fig. 2.3. Taking the natural logarithm of Eq. (2.6), we get

$$\ln N = \ln N_0 - \lambda t \tag{2.7}$$

Equation (2.7) shows that on a log scale the plot of N versus t is linear and $-\lambda$ is the slope of the straight line. Thus such a plot (Fig. 2.4) enables a ready determination of the decay constant for a nucleus of intermediate half-life.

The time in which the number of nuclei is reduced by half ($N = N_0/2$) can be obtained from Eq. (2.7) as

$$t_{\frac{1}{2}} = 0.693/\lambda \tag{2.8}$$

$t_{\frac{1}{2}}$ is called the half-life and is independent of the number of nuclei.

---

* The probability of electron capture depends on the density of electrons at the nucleus. The density can vary for different chemical compounds and therefore λ will be slightly different for different compounds.

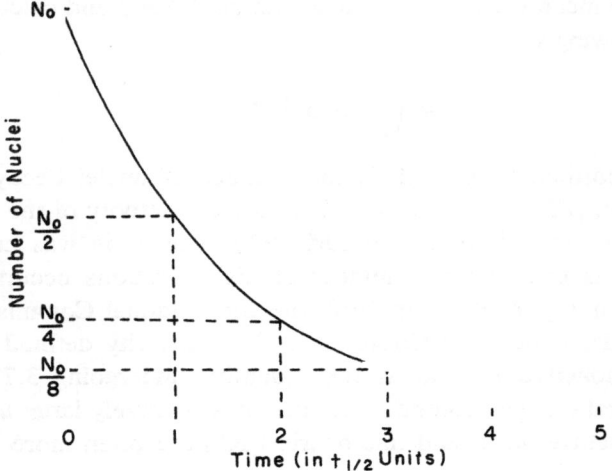

Figure 2-3. Exponential decrease of the number of radioactive nuclei with time.

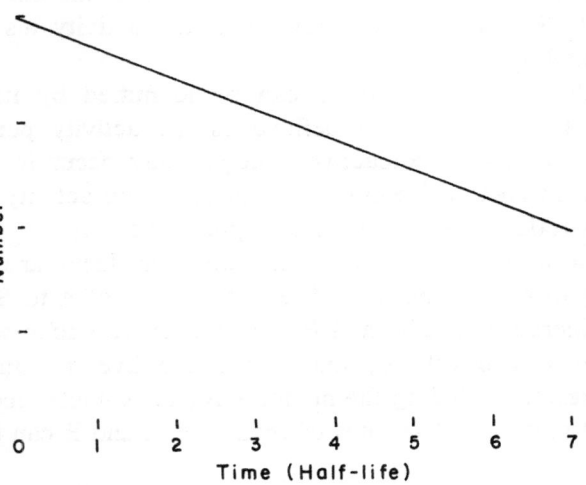

Figure 2-4. Number of radioactive nuclei (log scale) vs. time. The plot is a straight line.

The mean lifetime τ of nuclei can be defined and calculated in the following way:

$$\tau = \left(\int_{0}^{\infty} tN\,dt \right) \Big/ N_0 = 1/\lambda \qquad (2.9)$$

According to Eq. (2.5), the number of nuclei decaying per unit time, $|dN/dt|$, is λN. This is called the activity of the sample. The unit for activity is the curie (Ci). The curie was originally defined as equal to the number of disintegrations occurring per second in 1 g of ^{226}Ra. In 1950, the International Commission on Standards, Units, and Constants of Radioactivity defined 1 Ci of any radioactive nuclide as that quantity undergoing 3.7×10^{10} disintegrations per second. The curie is a relatively large unit, and the millicurie (mCi) and microcurie (μCi) are often more convenient.

Multiplying both sides of Eq. (2.6) by λ, we get

$$\lambda N = \lambda N_0 e^{-\lambda t}$$

or $\qquad\qquad\qquad\qquad\qquad\qquad\qquad\qquad\qquad\qquad (2.10)$

$$A = A_0 e^{-\lambda t}$$

where A is the activity at any time t and A_0 is the initial activity at $t = 0$. Equation (2.10) shows that the activity also changes exponentially.

Each radioactive species can be identified by its intrinsic specific activity, which is defined as the activity per gram of isotope. Normally, radioactive isotopes may occur in a mixture with nonradioactive isotopes. In this case, the activity per gram of the material is designated as the specific activity.

In a number of cases radioactive products are unstable, decaying to another nucleus. The series decay of A to B to C will be considered, where A and B are radioactive nuclei with decay constants respectively λ_A and λ_B and half lives $t_{A\frac{1}{2}}$ and $t_{B\frac{1}{2}}$. We are interested in finding the numbers N_A of A nuclei and N_B of B nuclei at any time. The rates of change of A and B can be written as

$$dN_A/dt = -\lambda_A N_A \qquad (2.11)$$

$$dN_B/dt = \lambda_A N_A - \lambda_B N_B \qquad (2.12)$$

The change in the number of B nuclei depends on the supply of B due to the decay of A nuclei and on the decay of B. If the number of A nuclei at $t = 0$ is N_{A_0}, we have

$$N_A = N_{A_0}e^{-\lambda_A t} \tag{2.13}$$

From (2.12) we obtain, if $N_B = 0$ at $t = 0$,

$$N_B = \frac{\lambda_A}{\lambda_B - \lambda_A} N_{A_0}(e^{-\lambda_A t} - e^{-\lambda_B t}) \tag{2.14}$$

2.3. CALCULATION OF ACTIVITIES PRODUCED BY NUCLEAR REACTIONS

Of the more than 1000 known radioactive nuclei, only about 50 occur naturally; the rest are produced artificially by nuclear reactions. In 1934, Irene Joliot-Curie and F. Joliot found that the bombarding of aluminum with alpha particles yielded a radioactive product. The reaction can be represented by the equation

$$^{27}_{13}Al + {}^{4}_{2}He \rightarrow {}^{30}_{15}P + {}^{1}_{0}n$$

where $^{30}_{15}P$ is the artificially produced radioactive nucleus, which decays by positron emission,

$$^{30}_{15}P \rightarrow {}^{30}_{14}Si + {}_{+1}^{0}e + {}^{0}_{0}\nu$$

We shall determine an expression for the activity of radioactive nuclei arising in such a nuclear reaction when the weight of the irradiated sample and the number of incident particles bombarding the sample are known. To make this calculation, we have to understand the concept of cross section.

Consider N hard spheres each of radius r_1, located in a plane of area A (Fig. 2.5). Assume further that a particle of radius r_2 impinges upon this plane. We regard a collision as occurring if the center of the impinging particle falls within or on a circle of radius $r_1 + r_2$ and with its center coinciding with that of the target particle. Hence the chance of a collision between the target particles and one incident particle is $N\pi(r_1 + r_2)^2/A$. If there are n

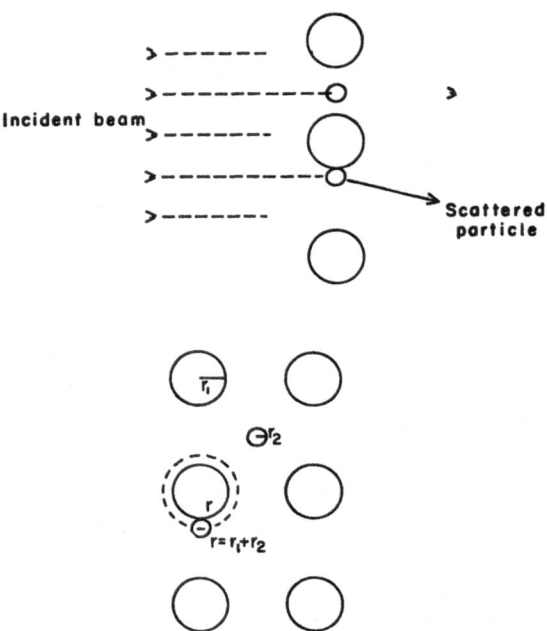

Figure 2-5. Elastic collision between incident particles of radius r_1 and target particles of radius r_2. A collision occurs when the center of the incident particle falls within an area of radius $r_i + r_2$ projected by each target particle.

particles per unit volume in the incident beam and they are traveling at a velocity v, the number of particles hitting the area A per second is nvA. The rate of collision R_t will be the number of incident particles times the chance of collision for each particle. It follows that

$$R_t = nvA \frac{N\pi(r_1 + r_2)^2}{A}$$

$$R_t = \phi N\sigma \qquad (2.15)$$

where $\phi = nv$, called the flux, is defined as the number of particles crossing a unit area per unit time. The quantity $\sigma = \pi(r_1 + r_2)^2$ is the area presented by a target particle for collision. If the center of the incident particle falls within this area, we have a collision. This area is called the reaction cross section.

For an elastic collision, the cross section is $\pi(r_1 + r_2)^2$. However, for nuclear reactions, the cross section is not necessarily the geometrical area. The cross section varies with the type of reaction and the energy of the incident particle. The interaction cross section may be smaller than, equal to, or larger than the geometrical cross section. The unit for the cross section is m^2 and the practical unit in general use is the barn ($= 10^{-28} \, m^2$).

With regard to the definition of flux as given in the foregoing, it is to be noted that this definition loses its meaning when we consider situations where the particles are not all traveling in one direction, but in all possible directions. In this case, the flux is defined, using Eq. (2.15), as

$$\phi = \frac{R_t}{N\sigma} \tag{2.16}$$

$$= \frac{\text{rate of reaction}}{\text{number of target nuclei} \times \text{reaction cross section}}$$

For example, the neutron flux inside a reactor is defined by Eq. (2.16).

We shall now derive expressions for the determination of the activity of the reaction products. Consider a reaction of the type

$$a + X \rightarrow B + c$$

where a is the incident particle, X is the target nucleus, and B and c are the reaction products, of which B is radioactive. The rate of change of the number of B nuclei N_B can be written as

$$dN_B/dt = R_t - N_B\lambda_B \tag{2.17}$$

where R_t is assumed constant since the change in the number of target nuclei is very small except in irradiations in high-flux reactors. If we assume $dR_t = 0$, Eq. (2.17) can be put in the form

$$\frac{d(R_t - \lambda_B N_B)}{R_t - \lambda_B N_B} = -\lambda_B \, dt \tag{2.18}$$

Integration of Eq. (2.18) yields

$$R_t - \lambda_B N_B = (R_t - \lambda_B N_B)_{t=0} e^{-\lambda_B t} \qquad (2.19)$$

Since at $t = 0$, $N_B = 0$, it follows from Eq. (2.19) that

$$N_B = (R_t/\lambda_B)(1 - e^{-\lambda_B t}) \qquad (2.20)$$

Figure 2.6 shows a plot of N_B versus t during and after irradiation. The number N_B decreases exponentially after irradiation. Also, it can be seen that it is not worthwhile to irradiate for more than two to three half-lives since $\frac{3}{4}$ to $\frac{7}{8}$ of the maximum number of radioactive nuclei (R_t/λ_B) have already been produced.

The activity during irradiation is given by

$$A_B = N_B \lambda_B = R_t(1 - e^{-\lambda_B t}) \qquad (2.21)$$

where R_t is related to the flux and cross section as given by Eq. (2.15). The activity A_B approaches a maximum when $t \to \infty$, that is,

$$A_{B\,max} = R_t \qquad (2.22)$$

Equation (2.22) was to be expected since N_B will be a maximum when the rate of decay is equal to the rate of production and

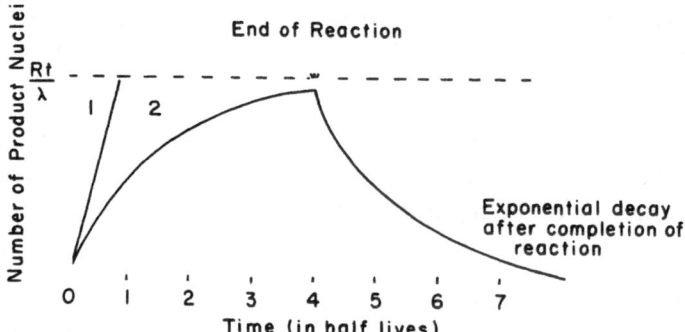

Figure 2-6. Change in number of radioactive product nuclei as a function of time during and after reaction (curve 2). If the product is not radioactive, the number increases continuously at a rate R as shown by curve 1.

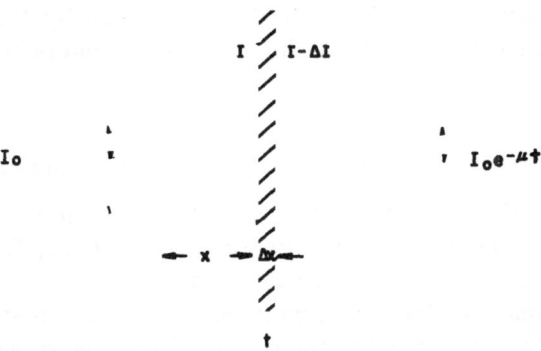

Figure 2-7. Variation of intensity as a function of the thickness of the absorber.

therefore the maximum activity will be equal to the rate of production.

At this point we can introduce a few ideas about absorption coefficients. Let us consider the case where we have N_v interacting centers in the target (nuclei, electrons, or any other interacting entity) per unit volume of the absorber. Let I denote the intensity,* that is, the number of incident particles crossing unit area per unit time at x (see Fig. 2.7). Then the change in I due to the passage through Δx of the target material is given by

$$\Delta I = -I\sigma N_v \, \Delta x \qquad (2.23)$$

The product σN_v is sometimes called the macroscopic cross section. This cross section is the rate of reaction per unit volume per incident particle. Integrating Eq. (2.23), we get†

$$I = I_0 e^{-\sigma N_v x} \qquad (2.24)$$

$$I = I_0 e^{-\mu_l x} \qquad (2.25)$$

where σN_v is replaced by μ_l, which is known as the linear

---

* The definition of flux ϕ is similar to the definition of I, the intensity. However, the term flux is used both for particles in a beam moving in one direction and for particles moving in random directions. The use of I is restricted to particles in a beam.

† Equation (2.25) is derived on the assumption that the particle is removed from the beam in one interaction. This is not true in the scattering of beta rays in air.

absorption coefficient. The quantity N_v is given by the following equations for the cases where the interacting centers are nuclei or electrons:

$$N_v = \rho N_A / A \qquad \text{for nuclear reactions}$$
$$N_v = \rho Z N_A / A \qquad \text{for electron scattering}$$

where Z is the atomic number of the target atoms and N_A is Avogadro's number. These equations apply for uniform targets containing atoms of a particular element.

Sometimes thickness is expressed in terms of areal density, g/cm^2. This can be done by making the following substitutions in Eq. (2.25):

$$\mu_l = \mu_m \rho, \qquad x = d_m / \rho$$

where d_m is the areal density. We then have

$$I = I_0 e^{-\mu_m d_m} \qquad (2.26)$$

where μ_m is called the mass absorption coefficient. The following relations hold for nuclear reactions:

$$\mu_m = \mu_l / \rho = \sigma N_v / \rho = \sigma \rho N_A / A \rho = \sigma N_A / A \qquad (2.27)$$

where N_A is Avogadro's number. A similar equation for electron scattering is

$$\mu_m = \sigma Z N_A / A \qquad (2.28)$$

From Eqs. (2.27) and (2.28), we see that μ_m is independent of the density, and, therefore, of the physical state of the absorber.

When the intensity decreases exponentially another useful quantity is the half-thickness $d_{\frac{1}{2}}$, defined by the equation

$$d_{\frac{1}{2}} = 0.693 / \mu_l \qquad (2.29)$$

The half-thickness is the thickness of material needed to reduce the intensity by half.

2.4. *INTERACTION OF HEAVY CHARGED PARTICLES WITH MATTER*

The interaction of charged particle radiation with matter can be considered as a series of collisions between the particle and the atoms in the medium. These collisions are either elastic, where the total kinetic energy of the colliding particle and the atom is conserved, or inelastic, where the kinetic energy is not conserved. In addition to transfer of energy, the incident particle undergoes a change in direction during collision. For the convenience of discussion, we will consider nuclear interaction and electronic interaction separately.

Elastic collision between the incident particle and primarily the nucleus is a Rutherford-type particle–nuclear scattering caused by electrostatic forces. The probability of scattering through a certain angle is proportional to the square of the charges of the projectile and the target. It is also inversely proportional to the energy. The differential cross section is inversely proportional to $\sin^4(\Theta/2)$ for a scattering into unit solid angle defined by $\Theta + d\Theta$. This means that the chance of large-angle scattering is much less than that of small-angle scattering. For example, the chance for backscattering is 10^{-5} times the chance for a $6°$ scattering.

The energy transfer in such collisions is proportional to the strength of the interaction and hence to the square of the charges. It is also inversely proportional to the mass of the nucleus. The maximum energy change occurs during a head-on collision, $180°$ scattering, and is given by

$$\frac{\text{energy of the projectile after collision}}{\text{energy of the projectile before collision}} = \left(\frac{M_2 - M_1}{M_2 + M_1}\right)^2$$

where M_1 and M_2 are masses of the projectile and the nucleus, respectively. One can see from this equation that if the masses are equal, the projectile gives all of its energy to the nucleus. The mass of the incident radiation is usually considerably less than that of the nucleus. Energy loss is therefore very small. Hence it can be concluded that the only effect of elastic nuclear scattering for alpha and beta particles is a change in direction.

The electrostatic force producing the deflection of the incident particle is also continuously changing the velocity of the particle, i.e., the particle is accelerating. The acceleration is proportional to the electrostatic force and it is inversely proportional to the mass M_1 of the particle; hence the acceleration is proportional to $Z_1 Z_2 e^2 / M_1$. Classical electromagnetic theory predicts the radiation of electromagnetic waves during an acceleration of the particle with an intensity proportional to the acceleration:

$$I_{\text{rad}} \propto Z_1^2 Z_2^2 e^4 / M_1^2 \qquad (2.30)$$

The intensity of the radiation is inversely proportional to the square of the mass and is directly proportional to the square of the charge of the nucleus. Energy loss by radiation is about one million times less for alpha particles than for electrons. Also, it is important to remember the Z_2^2 dependence in designing shielding and nuclear instrumentation. Electromagnetic waves thus produced are known as Bremsstrahlung. Even though classical theory predicts radiation during every acceleration, according to quantum theory, the emission of radiation occurs only in some cases. The probability for the emission of radiation is very small at low energies and increases with increasing energy.

Now we consider the elastic collisions between the incident radiation and the electron cloud of the atom. This interaction is the most important one for the discussion of the operation of detectors. The incident radiation attracts or repels the electrons. Through this electric force the particle supplies energy to the atomic electron; the amount of energy supplied varies due to the changing particle–electron distance. The energy thus gained takes the electron to one of its excited states or removes it completely from the atom. The excited electron falls back to the original level with the emission of a photon in a very short time ($\sim 10^{-12}$ sec). The electrons produced during ionization may possess enough kinetic energy to further ionize the atoms. This is called secondary ionization. Sometimes these particles have sufficiently long path lengths to make them visible in cloud chambers and they are called delta rays.

Particle–electron collisions also produce deflection of the projectile. Since the deflection probability is proportional to the square of the target particle, it is only $1/Z^2$ of the probability of deflection in a collision with a nucleus. The deflection angle depends on the mass difference of the colliding particles. The deflection is maximum when the masses are equal. The deflection during an alpha–electron collision is very small, while in β-ray-electron collision large-angle deflection may be expected. The alpha particles therefore travel in a straight-line path, while beta rays change their direction very often.

A charged particle in inelastic collisions with the atomic electrons loses energy and finally comes to a stop. The total distance a particle travels depends on the rate at which it loses energy in the medium. The energy loss per unit length $-dE/dx$ is called the stopping power. Let us consider the interaction of a radiation of charge $Z_1 e$, mass M, and velocity v passing through a medium containing N atoms of atomic number Z_2 per unit volume. The stopping power is given by

$$-\frac{dE}{dx} = \frac{4\pi Z_1^2 e^4}{mv^2} NZ_2 \left[\ln \frac{2mv^2}{I} - \ln(1 - \beta^2) - \beta^2 \right] \quad (2.31)$$

where m is the mass of the electron, $\beta = v/c$, and I is the geometric mean of the excitation and ionization potentials of the absorber atoms. Determination of I is usually made experimen-

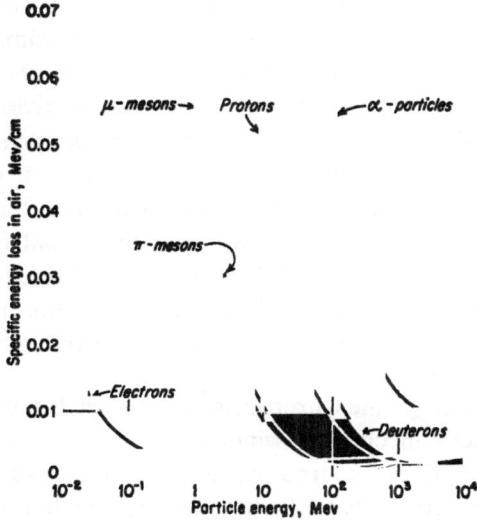

Figure 2-8. Specific energy loss in air for different particles. [From: A. Beiser, *Revs. Mod. Phys.* **24**, 273 (1952).]

tally; its theoretical estimation is a very complex problem. The value of I ranges from $10Z$ eV for heavy elements to $15Z$ eV for the light ones. A weak dependence on chemical binding has also been observed. However, since the dependence of I is logarithmic, these differences do not greatly affect the stopping power. Equation (2.31) is not valid for energies below 0.1 MeV, where the velocity of the particle is closer to the velocity of the atomic electrons. The energy lost per unit length depends on the charge of the incident particle and its velocity but not on its mass. It has been shown experimentally that the energy change per unit length is the same for protons, deuterons, and mesons of the same velocity.

The stopping powers for different particles in air are given in Fig. 2.8. For every particle the energy loss first decreases and then increases as the velocity of the particle approaches relativistic values. Physically this happens because the relativistic effect distorts the electric field for high velocities, reducing it in the direction along the path of the particle and increasing it in the perpendicular direction.

The product of energy loss per unit length [Eq. (2.31)] and the total energy E $(=\frac{1}{2}Mv^2)$, to a first approximation, is proportional to the mass and to the square of the charge of the incident radiation. Experimental systems for particle identification have been designed based on this fact. Such a system consists of a thin detector of known thickness followed by a thick detector which stops the particle. The signal from the thin detector is proportional to the stopping power and the signal from the thick detector is proportional to E. A product is obtained electronically and it is used for particle identification. This technique has become a routine one with the availability of thin, totally depleted semiconductor counters.

During energy measurements of charged particles, energy losses can occur in counter windows or other nonactive parts of the counter and in source covers for encapsulated sources. Therefore corrections have to be made for such energy losses. If the thicknesses of these layers are known, one can use Eq. (2.31) to make such corrections.

The detectable products of the energy loss are ions. Counting the number of ions is one way of finding out the energy expended by the radiation. However, the difficulty is that part of the energy is expended in exciting the electrons without ionization and most of the time the radiation expends more energy than required for ionization (such as during production of δ-rays, i.e., high-energy electrons capable of ionization). Therefore we use an average energy W required to produce an ion pair. This energy in a given material is practically independent of the kinetic energy or of the nature of the particle. The number of ions produced per unit length therefore is given by

$$-dE/dx = Wn_i \qquad (2.32)$$

Since W is constant, a plot of n_i vs. E will be similar to the curves in Fig. 2.8.

In the rest of this section we limit our discussion to the interaction of alpha particles. Figure 2.9 shows the tracks of alpha particles as seen in a cloud chamber. As mentioned earlier, because of the large mass difference between the alpha particle and the electron, energy loss in each collision is very small and the deflection is negligible. The alpha particle, which loses energy in a large number of collisions with electrons, therefore has a straight-line path. When the alpha particle has slowed down so that its velocity is lower than the velocity of the atomic electrons,

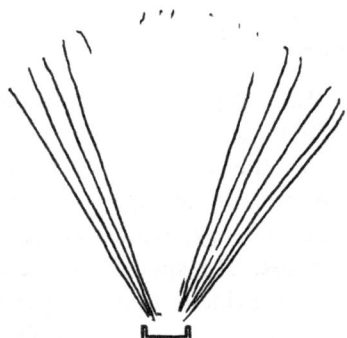

Figure 2-9. Alpha-particle tracks. Tracks are almost straight lines. Large-angle scattering with nuclei in air occurs very rarely.

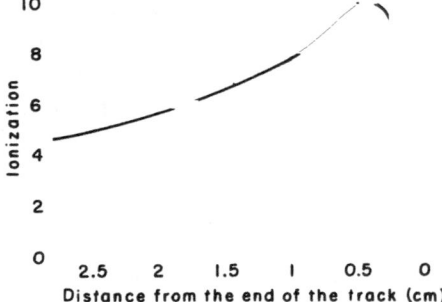

Figure 2-10. Specific ionization
of an alpha particle as a function
of the distance from the end of
the track. [From: M. G. Hollo-
way, and M. S. Livingston,
Phys. Rev. **54**, 18 (1938).]

it captures electrons and moves around as an ionized or neutral
atom. The collision from then on is atomic, not with electrons,
and the particle begins to "straggle."

For alpha particles, energy losses by Bremsstrahlung and
nuclear collision are negligible. Energy loss occurs mainly in
inelastic collisions with electrons. However, Eq. (2.31) has some
limitations: (1) For energies below 1.2 MeV we have an effective
charge of less than 2 because of the continuous capture and
release of the electrons. (2) The inner electrons do not participate
in collisions. Equation (2.31) is derived without considering these
factors and therefore a correction term is usually added to it. The
equation including this correction term is

$$-\frac{dE}{dx} = \frac{4\pi N Z_2 Z_1{}^2 e^4}{m_0 v^2} \left(\ln \frac{Z m_0 v^2}{I} + \ln \frac{1}{1 - \beta^2} - \beta^2 - \frac{C}{Z} \right) \quad (2.33)$$

The contribution to the energy loss by this term C/Z is less than
1% and it has been calculated recently.[1]

Figure 2.10 shows the variation of specific ionization with
distance from the source for an alpha particle. Near the source
the particle has a high energy, which decreases as the distance
from the source increases. The specific ionization increases as the
distance increases. It is evident that, as the velocity decreases,
the time of interaction (time spent by the alpha particle near the
atom) increases and therefore more ions are produced. The
ionization at the end of the track decreases very rapidly because

the effective charge of the alpha particle decreases due to capture and release of electrons by it.

The range of an ionizing particle can be considered as the distance from the source to the point where the last ion is produced. Of course this distance will not always be the same for all particles of the same type and the same energy. The variation occurs because of the possible variations in the specific ionization and the variations in the energy required for ionization. Therefore we define a mean range

$$\bar{R} = \int_0^{\bar{R}} dx = \int_0^{E_0} (-dE/dx)^{-1} \, dE$$

The mean range $\bar{R}$ cannot be calculated by substituting for dE/dx from Eq. (2.31), because of the inapplicability of the equation for very low energies. Empirical equations for mean range for different particles have been obtained. The empirical equation for alpha particles in air as a function of energy in MeV is (R in cm)

$$R = 0.318E^{3/2} \tag{2.34}$$

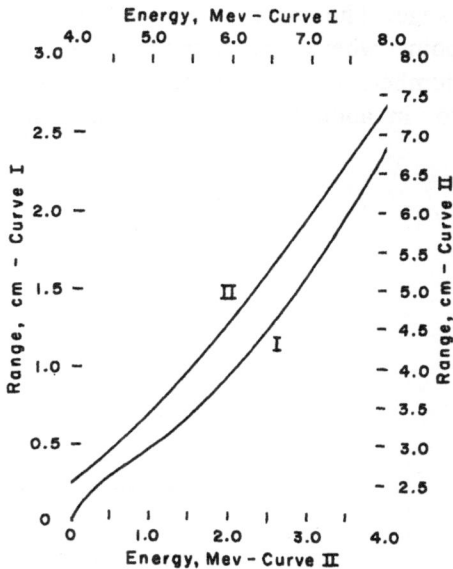

Figure 2-11. Range-energy relationship for low-energy alpha particles in air at 115°C and 760 Torr. [From: H. A. Bethe, *Revs. Mod. Phys.* **22**, 213 (1950).]

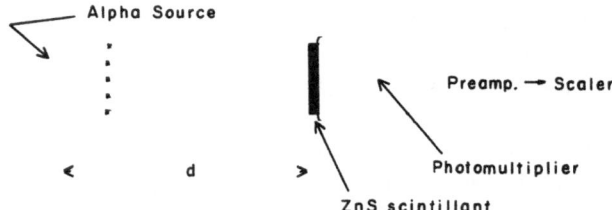

Figure 2-12.   Experimental arrangement for the determination
of the range of alpha particles in air.

This relation, known as the Geiger relation, is good only for a limited energy range. At low energies $R \propto E^{3/4}$, while at high energies the dependence is more nearly $R \propto E^2$. The range-energy relationship is shown in Fig. 2.11.

The range of particles can be determined by the simple experimental arrangement shown in Fig. 2.12. The variation of the number of counts for different source–counter distances is shown in Fig. 2.13. The section of the curve BC is almost linear and extrapolation of BC toward the axis gives the extrapolated range. The mean range is the distance from the source to the point where the number of counts is half of the maximum number. The tailing off at D is known as straggling. This is due to, in addition to the reasons mentioned before, the statistical

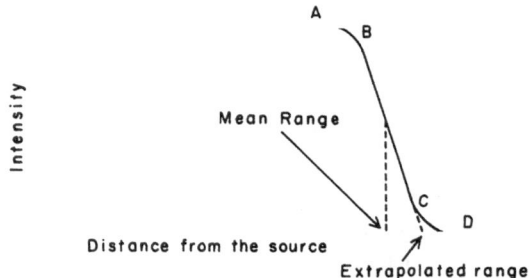

Figure 2-13.   Variation of the number of alpha particles
as a function of source–counter distance.

nature of the collision process, both in the number of electron–ion pairs and in the amount of energy transferred in each collision.

For high-energy radiation, ranges are determined in materials other than air. An approximate expression for the alpha-particle range in solids is

$$R_s = \frac{3.2 \times 10^{-4} A^{\frac{1}{3}}}{\rho} R \tag{2.35}$$

where R_s is the range in the solid (g/cm²), R the range in air in cm, ρ the density of the solid (g/cm³), and A the mass number. Biologists are interested in finding the range in tissue. A simple relationship for this is

$$R_{air}\rho_{air} = R_{tissue}\rho_{tissue} \tag{2.36}$$

A relative stopping power can also be defined as

$$RSP = \frac{\text{range in air}}{\text{range in the material}}$$

where RSP is a function of the velocity of the particle. Sometimes the stopping power of a material is expressed in terms of equivalent thickness in units of mg/cm², defined as (g/cm²) × 1000. A related quantity, thickness of the material equivalent to 1 cm of air, can be defined as thickness in mg/cm² equivalent to 1 cm of air = density × 1000/RSP. Values of the above quantities for commonly used absorbers for alpha particles from ^{214}Po (RaC′) with an extrapolated range of 6.953 cm in air at 15°C and 760 Torr are given in Table 2.1. As can be seen from the table, very thin sheets of Al or mica will stop all the alpha particles from radioactive substances. Even a single sheet of paper is enough to stop alpha particles. One has to be mindful of this while selecting windows for counters. Information on stopping power and ranges is also necessary in choosing the thickness of the active volume of a counter.

2.5. ELECTRON INTERACTION

Figures 2.10 and 2.14 show the main differences between alpha particles and electrons in their interactions with matter. Specific

Table 2.1. Range and Stopping Power for ^{214}Po Alpha Particles in Various Substances[a]

Substance	Extrapolated range, cm	RSP	ρ, g/cm^3	Equivalent thickness, mg/cm^2	Thickness in mg/cm^2 equivalent to 1 cm of air
Mica	0.0036	1930	2.8	10.1	1.45
Aluminum	0.00406	1700	2.70	11.0	1.57
Copper	0.00183	3800	8.93	16.3	2.35
Gold	0.00140	4950	19.33	27.1	3.89

[a] Data taken from Ref. 4.

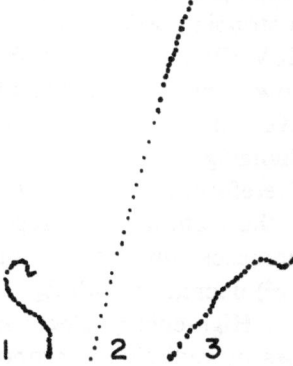

Figure 2-14. Tracks of electrons. Low-energy electrons suffer a direction change more often than the high-energy electron. Also note the increase in ionization as the energy decreases.

ionization is very low for electrons and the path length is longer than for alpha particles. Thus a 3-MeV electron produces only 4 ions/mm while for the same energy an alpha particle produces about 4000 ions/mm. The path length is 1000 cm for electrons and 2.8 cm for alpha particles. The paths of alpha rays are straight, scattering by nuclei (Rutherford scattering) occurring very rarely. The path of an electron is not straight, because the electron suffers a large amount of scattering by atomic electrons and nuclei. In the case of electrons, the identity of the incident and scattered particles causes further complications. Almost complete transfer of energy is possible in a collision. Therefore, after an electron–electron collision, the electron with the higher energy is considered the incident particle. Also, there is no change in the effective charge and therefore the energy loss of electrons by ionization is appreciable down to energies in the eV range.

Another important point to be considered is that even for low energies the electron velocity is high enough to produce relativistic effects. Bethe[2] derived the following expression for energy loss by ionization:

$$\left(-\frac{dE}{dx}\right)_i = \frac{2\pi e^4 N Z_2}{m_0 v^2} \left\{ \ln \frac{E(E + m_0 c^2)^2 \beta^2}{2I^2 m_0 c^2} \right.$$
$$- \ln 2[2(1 - \beta^2)^{\frac{1}{2}} - 1 + \beta^2]$$
$$\left. + (1 - \beta^2) + 1/8[1 - (1 - \beta^2)^{\frac{1}{2}}]^2 \right\} \quad (2.37)$$

where $\beta = v/c$. The effect of the correction factors is an increase in stopping power, after reaching a minimum, for energies above 1 MeV (Fig. 2.8). Since E is in the log term, this increase is very slow. The dependence of the stopping power on the material is given by NZ_2 (= $\rho N_A Z_2/A$). Variation of Z_2/A is very slow, changing from 0.5 for low-Z materials to 0.39 for uranium. Therefore, for a given E, $(dE/dx)_i$ mainly depends on the density of the material. The dependence of I on Z_2 cannot be neglected. I increases with atomic number and therefore dE/dx, in MeV/(g/cm^2) decreases with Z_2.

High-energy electrons also lose energy by radiation. Energy loss by radiation is proportional to Z^2, the square of the charge of the nucleus in the medium, and increases linearly with energy, while the energy loss by ionization is proportional to Z and logarithmically proportional to energy. The ratio of the energy losses due to these different processes is given by

$$\frac{(dE/dx)_{\text{rad}}}{(dE/dx)_{\text{ion}}} = \frac{EZ}{800} \qquad (2.38)$$

Energy loss by radiation in lead, $Z = 82$, is significant, $\sim 10\%$ even at $E = 1$ MeV. Figure 2.15 shows the energy loss in lead due to both processes.

For electrons the range is not well defined compared to alpha particles. The reasons for this are that (1) electrons suffer changes in direction by collisions, and (2) straggling is much greater for electrons than for alpha particles, because of the large variations in energy transfer. For electrons from radioactive sources, there is an additional reason, namely the continuous energy distribution of the electrons from radioactive sources.

Since the path length is very long for electrons in air, metal foils, commonly Al foil, are used for absorption experiments. For monoenergetic electrons the number of electrons decreases almost linearly with thickness of the absorber and for electrons from radioactive sources the variation is almost exponential. A "visual" method to determine range is to find the point where the absorption curve meets the background. A useful empirical range–

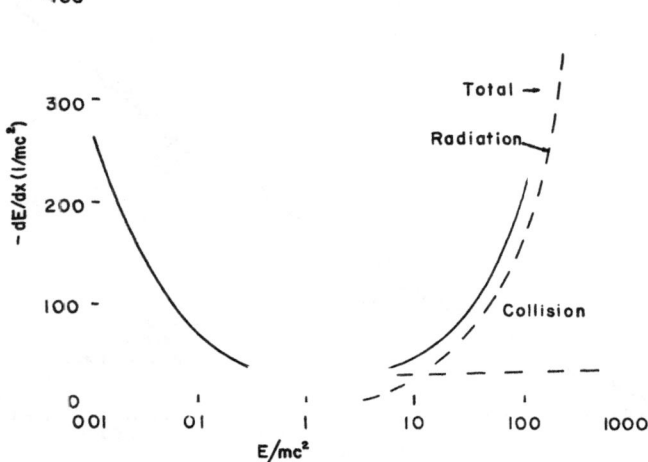

Figure 2-15. Energy loss rate for electrons by collision and radiation in lead as a function of energy. [By permission, from: W. Heitler, *The Quantum Theory of Radiation*, Oxford University Press, New York, 1944.]

energy relation is

$$R = 412E^n \tag{2.39}$$

where $n = 1.265 - 0.095 \ln E$ for $0.01 < E < 3$ MeV. Note that in this energy range, n is not a constant but has an energy dependence. For higher energies a more suitable equation is

$$R = 530E - 106 \quad \text{for} \quad 1 < E < 20 \text{ MeV}$$

In the above equations, the range R is in mg/cm^2 and E is in MeV. The range–energy relationship is shown in Fig. 2.16.

An exponential decrease in intensity with thickness is obtained when a single interaction (collision) removes the particle from the beam. This does not happen in electron collisions. However, a combination of factors, including the continuous energy distribution of the electrons, the large number of collisions involved, and the change in direction during each collision,

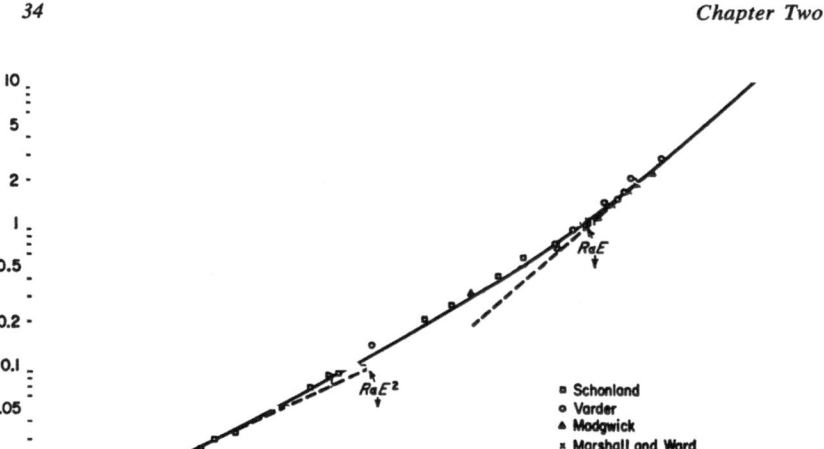

Figure 2-16.   Range-energy relationship of beta rays in aluminum. [From: R. D.
Evan, *The Atomic Nucleus*, p. 624, McGraw-Hill, New York, 1955.]

produces the effect of an exponential intensity decrease. Therefore one gets for intensity

$$I = I_0 e^{-\mu_m d_m} \tag{2.40}$$

where μ_m is the material absorption coefficient and d_m is the areal
density.

An empirical relation for the absorption coefficient is

$$\mu_m = 17/E_m^{1.14} \tag{2.41}$$

where E_m is the maximum energy of the beta ray. This
approximate expression gives sufficiently good values for μ_m in
the range $0.1 < E_m < 4$ MeV.

2.6.   BACKSCATTERING OF ELECTRONS

As mentioned before, electrons suffer large-angle scattering more
often than heavy particles, such as alpha particles. This scattering
is to a considerable degree due to the electric field of the nucleus.

The large-angle scattering, backscattering, is significant in our discussion in two respects: (1) In an experimental arrangement such as shown in Fig. 2.17 the number of radiations reaching the counter will depend on the backing of the sample. Consequently, there will be more counts than for the simple solid geometry calculation for a radioactive source without backing. (2) When electrons strike a counter (in particular, a solid counter), some of them will be scattered back. This will effectively reduce the efficiency of the counter. (3) Scattered beta particles reaching the detector will contribute heavily to the low-energy side, distorting the beta spectrum. In experiments for accurate determination of the shape of the beta-spectrum special care should be taken to reduce the effect of backscattered beta particles.

Figure 2.18 shows a typical result of an experiment when the counts, number of beta particles reaching the counter, increases with the thickness of the backing material of the source, eventually attaining a maximum value, known as the "saturation" value. The ratio of the counts with backscatter to the counts without backscatter is called the backscattering factor f_{bs}. The ratio where the thickness of the backscatterer is sufficient to produce saturation backscattering, designated $(f_{bs})_{sat}$, depends on the material of the scatterer and increases with the Z value of the material. Dangui[3] has shown that the saturation scattering factor is proportional to

$$[Z(Z + 1)/M_2]^{\frac{1}{2}} \qquad (2.42)$$

His results are shown in Fig. 2.19. No energy dependence for $(f_{bs})_{sat}$ has been observed. The thickness required for saturation

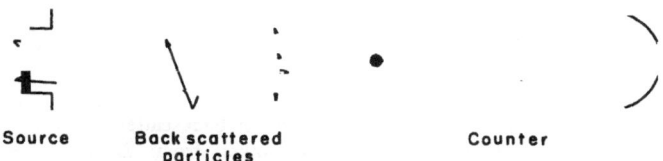

Source   Back scattered   Counter
particles

Figure 2-17. Experimental setup for β-counting, showing the effect of backscattering.

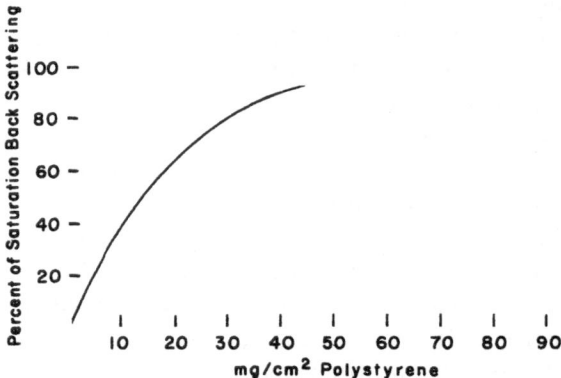

Figure 2-18. Backscattering growth curve for RaE radiation (1.17 MeV), polystyrene backing. [Data taken from: L. R. Zumwalt, U.S. Atomic Energy Commission Report AECU-567, 1950.]

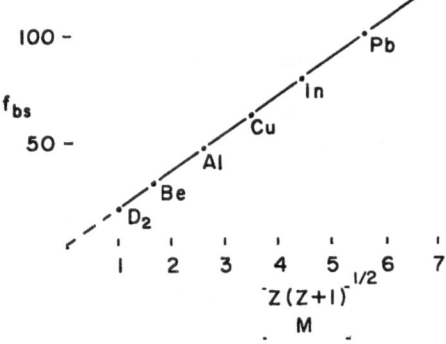

Figure 2-19. Variation of saturation backscattering as a function of Z. [Data taken from: L. Dangui, Institut Interuniversitaire des Sciences Nucleaires, Monographie No. 10, Brussels.]

backscattering is found to be about 25% of the range of the particles in the material.

From Eq. (2.42) one can see that the number of beta particles backscattered increases with Z, and therefore to reduce the number scattered back from solid state counters, they should be of low-Z materials. For example, backscattering will be less from an organic counter than from an inorganic one such as a NaI(Tl) scintillant. Similarly, among the semiconductor counters, silicon counters will be preferable to germanium counters for electron detection because of the lower value of Z and, hence, lower backscattering.

2.7. INTERACTION OF GAMMA AND X RAYS WITH MATTER

Gamma rays are removed from a beam when it passes through a medium in a single process, either by scattering or absorption. Hence the intensity variation as a function of absorber thickness is given by Eq. (2.25):

$$I = I_0 e^{-\mu x}$$

The intensity decreases exponentially. As we have seen, alpha particles are stopped by many interactions involving practically no change in their direction. Therefore the absorption curve, the number of particles versus the thickness, is a line parallel to the x axis to the end of the range. Monoenergetic electrons are also stopped by many interactions but they suffer many small changes in their direction. However, they are very rarely removed from the beam in a single collision. The absorption curve for electrons is almost a straight line linearly decreasing with thickness, lying somewhat between the alpha and the gamma absorption curves. The absorption curve for beta rays from radioactive sources is exponential.

The absorption coefficient is a function of energy for any material. For monoenergetic gamma rays, the intensity decreases exponentially with thickness. If μ_l can be determined experimen-

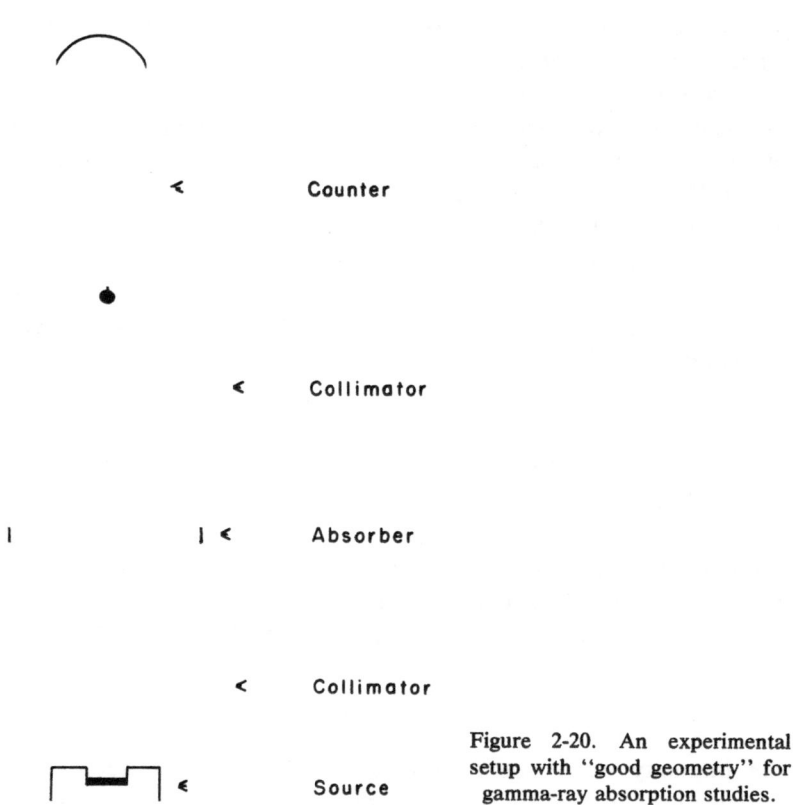

Figure 2-20. An experimental setup with "good geometry" for gamma-ray absorption studies.

tally, the energy of the gamma ray can be determined. Some care should be taken in designing such experiments. The experimental setup should have a "good geometry," i.e., the shielding materials and the counter should be properly located so that the scattered gamma rays do not reach the counter. Such an arrangement is shown in Fig. 2.20. The absorption method, even though it is capable of giving the energy of the gamma rays, has limited accuracy and cannot be used for gamma rays of more than one energy.

The absorption of gamma rays is due mainly to three different processes: the photoelectric effect, the Compton effect, and pair production. The absorption coefficient can therefore be written as

$$\mu_l = {}_{PE}\mu_l + {}_{C}\mu_l + {}_{PP}\mu_l \tag{2.43}$$

Contributions to μ_l due to other processes, such as Rayleigh scattering, gamma–nuclear reactions, and nuclear resonance absorption, are very small compared to the contributions of the three processes mentioned above.

2.7.1. *Photoelectric Effect*

In the photoelectric effect the incident photon knocks an electron out of the atom. Conservation of momentum and energy is expressed by the following equations:

$$p_\gamma = p_e + p_{atom} \tag{2.44}$$

$$E = T_e + T_{atom} + I \tag{2.45}$$

where I is the ionization energy, the energy required to remove the electron from the atom; and T_e and T_{atom} are the kinetic energies of the electron and the atom, respectively. Even though the atom acquires momentum in accordance with the law of conservation of momentum, the energy acquired by the atom is very small. The ratio of the kinetic energy of the electron to the kinetic energy of the atom is equal to the ratio of the mass of the atom to the mass of the electron, and is approximately equal to 10^4. Therefore Eq. (2.45) can be approximated by

$$h\nu = T_e + I \tag{2.46}$$

Since the atom is left with a vacancy in an atomic shell, it will be immediately filled by an outer electron and an x ray will be emitted.

When radiation is absorbed in a counter by the photoelectric effect, an electron and an x ray will be released, their total energy being equal to the energy of the incident radiation. If the electron is absorbed and the x ray escapes, the pulse produced by the counter (proportional type) will be smaller than the pulse produced in the case when both photoelectron and x ray are absorbed in the active volume. X-ray escape is significant in gas

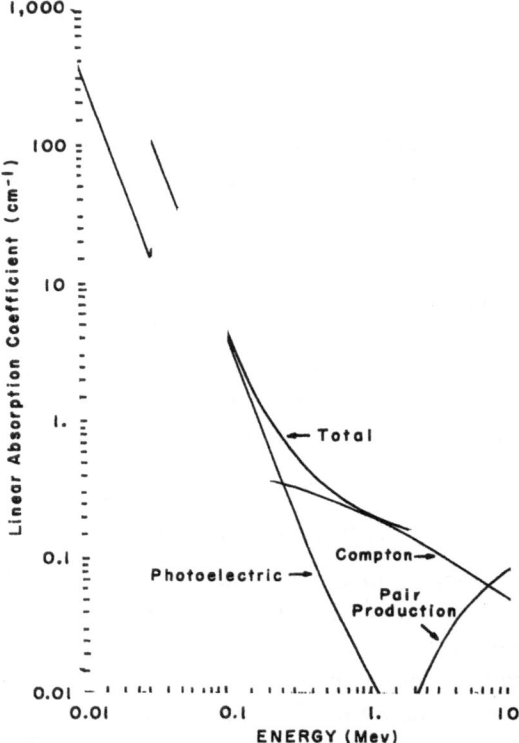

Figure 2-21. Variation of photoelectric, Compton,
pair production, and total cross sections for sodium
iodide as a function of energy.

counters and semiconductor counters. When proportional
counters are used to detect low-energy photons, one can get two
groups of pulses: one due to total absorption and the other due to
the escape of x rays.

The variation of the photoelectric cross section with energy is
shown in Fig. 2.21. In the low-energy region marked variations in
the cross section are evident. These occur at points where the
energy of the incident radiation is exactly equal to the ionization
energy of one of the electrons. These sudden drops, as the energy
decreases, are called absorption edges.

For gamma-ray energies above the ionization energy for K electrons, the atomic cross section is given in barns by

$$_{PE}\sigma_{A(K)} = 40S(Z - 0.3)^5(h\nu)^{-7/2} \qquad (2.47)$$

where S is a complicated function of energy and atomic number. From Eq. (2.47) it can be seen that the cross section is proportional to the fifth power of the atomic number of the absorbing material and is inversely proportional to $(h\nu)^{7/2}$. For low-Z materials, such as biological tissues, the photoelectric cross section is negligible for energies above 200 keV, while for high-Z materials like lead it is appreciable even for 1–2 MeV energies.

Equation (2.47) is only approximate and it considers only K electrons. The effect of including scattering by other electrons is a change in the exponent of Z. The exponent of Z at 1.13 MeV is approximately 4.5; at 2.62 MeV it is 4.6. The cross section varies with energy as $(h\nu)^{-3}$ for energies less than 0.5 MeV, $(h\nu)^{-1}$ for energies greater than 1 MeV, and as $(h\nu)^{-2}$ between these energy values.

The angular distribution of scattered electrons depends on the energy of the gamma rays. For low energies 90° scattering is favored. This is to be expected since the electric vector is perpendicular to the direction of the radiation. As the energy increases, the angle of scattering becomes smaller.

2.7.2. *Compton Effect*

In the Compton effect the photon undergoes elastic scattering. Part of the energy of the incident photon is carried away by the electron and the rest appears in the form of a photon of higher wavelength. As in any elastic scattering, we can write the equations for the conservation of momentum and energy:

$$h\nu = h\nu' + T_e \qquad (2.48)$$

$$(h/c)\nu = (h/c)\nu' + p_e \cos \theta \qquad (2.49)$$

$$0 = (h/c)\nu' \sin \theta + p_e \sin \phi \qquad (2.50)$$

where, as indicated in Fig. 2.22, $h\nu$ and $h\nu'$ are the energies of the

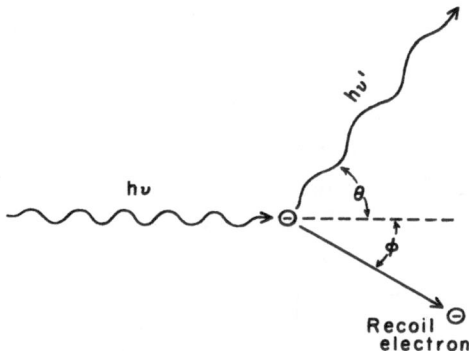

Figure 2-22. The collision between a photon
and an electron in Compton scattering.

photon before and after collision; T_e and p_e are the kinetic energy
and momentum of the electron after scattering; and θ and ϕ are
the angles of scattering of the photon and the electron. Solving
the above equations, we get an equation for the change in
wavelength

$$\Delta\lambda = \frac{c}{\nu} - \frac{c}{\nu'} = \frac{h}{m_0 c}(1 - \cos\theta) \qquad (2.51)$$

Expressed in angstroms, the change in wavelength is

$$\Delta\lambda = 0.0242(1 - \cos\theta) \qquad (2.52)$$

From Eq. (2.52), it can be seen that the change in wavelength of
the scattered photon depends on the angle and not on the energy.
The fractional change in wavelength $\Delta\lambda/\lambda$ will be higher for high-
energy photons scattered at the same angle.

The relation obtained by Klein and Nishina for the cross
section per electron $_c\sigma_e$ in the removal of photons from the
incident beam by Compton scattering is

$$_c\sigma_e = 3\sigma_0/4\alpha^2 \qquad (2.53)$$

where $\alpha = h\nu/m_0 c^2$ and σ_0 is the Thomson scattering cross
section. The linear absorption coefficient $_c\mu_l$ and the mass

absorption coefficient $_c\mu_m$ can be obtained from Eq. (2.28):

$$_c\mu_l(\text{cm}^{-1}) = \rho N(Z/A)\,_c\sigma_e$$

$$_c\mu_m(\text{cm}^2/\text{mg}) = N(Z/A)\,_c\sigma_e$$

For low-atomic-number nuclei, $A \simeq 2Z$, the mass absorption coefficient is independent of Z. Variation of the linear absorption coefficient with energy is shown in Fig. 2.21. The total scattering coefficient per electron decreases with energy. The decrease is quite slow for low energies and for energies above 0.5 MeV the decrease is proportional to $(h\nu)^{-1}$. Thus the decrease in the Compton cross section is much slower than the decrease in the photoelectric effect. For high-Z materials, Compton scattering is predominant in the energy range of 0.6–2.5 MeV.

The kinetic energy T_e of the scattered electrons is given by

$$T_e = h\nu \left[1 + \frac{m_0 c^2}{h\nu(1 - \cos\theta)} \right]^{-1} \tag{2.54}$$

and for $\theta = 180°$ (backscattering), T_e is a maximum given by

$$T_{e,\,\text{max}} = \frac{h\nu}{1 + (m_0 c^2/2h\nu)} \tag{2.55}$$

For high-energy gamma rays, $m_0 c^2 \ll 2h\nu$. In this case, the maximum kinetic energy of the scattered electrons can be obtained from $T_{e,\,\text{max}}(\text{MeV}) = h\nu - 0.255$.

2.7.3. Pair Production

When a gamma ray having energy greater than 1.02 MeV passes near the electric field of a nucleus, an electron–positron pair is created. The conservation of energy is satisfied in the following way:

$$h\nu = 2m_0 c^2 + T_e + T_p$$

where T_e and T_p are the kinetic energies of the electron and positron, respectively. The presence of the nucleus is necessary to conserve momentum. The energy taken by the nucleus is extremely small.

The equation for the variation of the cross section with energy is complicated. The cross section is proportional to Z^2 and increases with energy first slowly and then faster. For energies above 2 MeV this is the most important process of interaction (see Fig. 2.21). The cross section for pair production is approximately proportional to $\ln E$.

The positron and electron lose energy to the medium by Coulomb interaction. The positron, after it has slowed down, captures an electron and undergoes annihilation. The process is almost the reverse of pair production. In annihilation the rest-mass energies of the positron and electron appear in the form of two gamma rays, each with 0.51 MeV energy. The two gamma rays go in opposite directions, thereby conserving overall momentum.

2.8. *INTERACTION OF NEUTRONS*

Neutrons are neutral particles existing in nuclei. These particles were discovered by Chadwick in 1932. They have no electric charge and their mass is slightly higher than the mass of the proton: $m(^1_0n) = 1.0086654$ amu, whereas $m(^1_1p) = 1.007276$ amu.* Free neutrons decay to protons with a half-life of 12 ± 1.5 min.

Neutrons are released in nuclear reactions and in the fission process. Two typical neutron-producing reactions are

$$^4_2He + \,^9_4Be \rightarrow \,^{12}_6C + \,^1_0n \qquad (2.56)$$

$$^2_1H + \,^3_1H \rightarrow \,^4_2He + \,^1_0n \qquad (2.57)$$

The reaction indicated by Eq. (2.56) is important in low-flux neutron sources. A common source of neutrons is $^{210}_{84}Po$ surrounded by a cover of ^9_4Be. Neutrons are produced when the alpha particles from the polonium nuclei react with the beryllium

---

* Based on ^{12}C mass unit = 12.00000 amu, i.e., the atomic mass unit is 1/12 of the mass of ^{12}C.

nuclei. Such sources could be prepared with any alpha-active source, such as $^{226}_{88}$Ra, $^{239}_{94}$Pu, and $^{241}_{95}$Am. This type of source yields about 10^6 neutrons/sec per Ci of alpha source.

Fission is another source of neutrons. About 2.5 neutrons are produced in every fission event. In reactors where fission is used to produce power, neutron fluxes of 10^{11}–10^{15} n/cm^2-sec are very common. The isotope ^{252}Cf is a man-made isotope which undergoes spontaneous fission, emitting a large number of neutrons (2.34×10^{12} n/sec-g) with a sufficiently long half-life. This isotope therefore can be used as a neutron source. Such a source has the advantages of lower heat generation and smaller volume than alpha-beryllium sources.

Neutrons interact with matter mainly through two processes: elastic collisions and nuclear reactions. In elastic collisions energy loss depends on the angle of collision. The maximum energy after collision is the energy of the incident neutron T and the minimum energy is given by

$$T_{\min} = T \left(\frac{M_1 - M_2}{M_1 + M_2}\right)^2$$

where M_1 is the mass of the nucleus and M_2 is the mass of the neutron. Neutron–proton elastic collisions are used for the detection of neutrons. The neutron is allowed to strike a thin foil of hydrogenous material on the wall of the counter. Protons will then emerge from the foil with energy between zero and T.

We are interested in those neutron-induced reactions that produce charged particles. The most commonly used reactions are

$$^{1}_{0}n + {}^{10}B \rightarrow {}^{7}Li + {}_{2}^{4}He + 2.8 \text{ MeV}$$

$$^{1}_{0}n + {}^{6}_{3}Li \rightarrow {}_{2}^{4}He + {}_{1}^{3}H + 4.78 \text{ Me}$$

$$^{1}_{0}n + {}^{235}U \rightarrow \text{fission} + 200 \text{ MeV}$$

All these reactions have high cross sections for low-energy neutrons and are therefore especially suitable for the detection of thermal neutrons. The counters are usually designed such that the reaction products lose all their energy in the active volume of the counter.

Reactions where the products are radioactive are sometimes

used for the detection of neutrons. Foils of materials of known reaction cross section are activated in the neutron beam, and the neutron flux can be determined from the measurement of induced activity.

2.9. COSMIC RAYS

Cosmic rays reaching the surface of the earth are made up of two components, called the soft and hard components. The soft component is the part that can be easily absorbed. It is composed of charged heavy particles, such as mesons, protons, etc. The hard component is made up of electrons, gamma rays, and neutrons. The significance of these radiations in standard radiation laboratories arises from their contribution to the background.

REFERENCES

1. "Studies in Penetration of Charged Particles in Matter," NAS–NRC Report 1133, Washington, D.C., 1933.
2. H. A. Bethe, *Revs. Mod. Phys.* **22,** 213 (1950).
3. L. Dangui, Instituto Interuniversitaire des Sciences Nucleaires, Monographie No. 10, Brussels.
4. I. Kaplan, *Nuclear Physics*, Addison-Wesley, Reading, Massachusetts.

BIBLIOGRAPHY

R. D. Evans, *The Atomic Nucleus*, McGraw-Hill, New York, 1955, Chapters 18–25.
K. Siegbahn, ed., *Alpha- Beta-, and Gamma-Ray Spectroscopy*, North-Holland Publishing Co., Amsterdam, 1965, Chapters I and II.

Gas Counters

Gas counters for the detection of radiation were developed in 1908 by Geiger in Rutherford's laboratory. These counters became practical for the measurement of radiation shortly thereafter, even though scintillants were for a long time in use for the detection of radiation. The original Geiger counter was similar in construction to that of Fig. 3.1. It had an outer cylindrical electrode and an inner wire electrode with a potential difference applied between them. Geiger found that, when a radiation was stopped in the counter, a current flowed through the electrodes which was detectable with an electrometer of "moderate sensibility."

3.1. *GENERAL PROPERTIES OF GAS COUNTERS*

To understand the working of a counter, we have to understand the behavior of ions produced in the counter by the incident radiation. Let us assume that radiation of 1 MeV is stopped in the counter. On the average, the incident radiation expends about 30 eV to produce an ion pair, that is, a positive ion and an electron. The number n of primary ion pairs produced by this radiation is

$$n = 10^6 \text{ eV}/30 \text{ eV} = 3.3 \times 10^4$$

Since we have electrons and positive ions, their behavior will

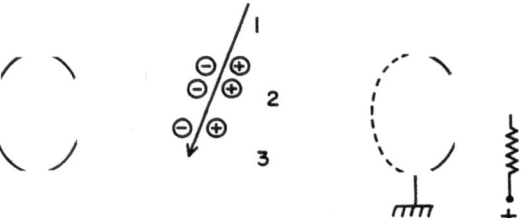

Figure 3-1. Geiger counter. A similar cylindrical geometry is used in most gas counters; 1, incident radiation; 2, central electrode; 3, outer electrode.

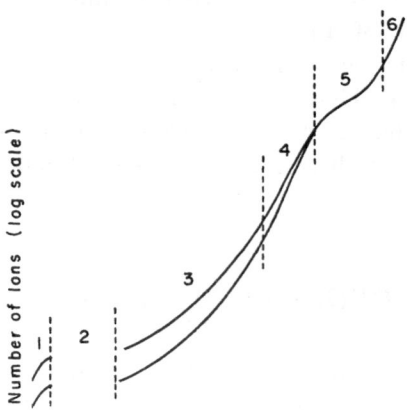

Applied Voltage

Figure 3-2. Variation of the number of ions collected by the electrodes as a function of the applied voltage. The upper curve is for a 2-MeV and the lower curve for a 1-MeV radiation. The different regions are: (1) recombination, (2) ionization, (3) proportionality, (4) limited proportionality, (5) Geiger, and (6) discharge region.

depend on the electric field inside the counter. If there is no field, the ions will recombine. As the applied voltage increases, more and more positive ions and electrons will be collected by the electrodes. The electrons will be collected by the central electrode (which is normally positive) and the positive ions will be collected by the outer electrode. After a certain voltage is reached, all the primary ions, that is, those ions and electrons resulting directly from the incident radiation, are collected. A plot of the number of ions collected by the electrode versus the applied voltage is shown in Fig. 3.2. In the first region of this figure, called the recombination region, the voltage is insufficient to collect all the ions, which may simply recombine. In the second region, called the ionization region, the electric field is sufficiently large to sweep all the ions to the electrodes. A counter operating in this region is called an ionization counter.

The ions travel to the electrodes with an average velocity known as the drift velocity. This velocity depends on the applied field, pressure of the gas, mass of the ion, geometric size of the counter, etc. The drift velocity for the electron is $\sim 10^3$ times the drift velocity for a positive ion.* During the transit of the electrons or positive ions, they collide with other ions and atoms in the counter. We can also think of the mean free path for the ions as the average distance an ion travels between two collisions. With increasing electric field, an ion can gain enough energy in one mean free path to ionize the next atom with which it collides.

---

* The drift velocity of the electrons, the average velocity they gain in the direction of the field between two collisions, is much higher than that of the positive ions. The two main reasons for this are the differences in their mean free paths and masses. The mean free path of the electrons is larger than that of the positive ions because of the difference in their radii, and the mass of the electron is less than that of the ions.

The drift velocity w for heavy ions is given by

$$w = \mu E/P$$

where μ is the mobility, E is the electric field, and P is the pressure of the gas. μ_+, the mobility for the positive argon ion, is 1040 (cm/sec)(V/cm)$^{-1}$(Torr), while μ_-, for the negative argon ion, is 1290. Unfortunately, there is no linear relationship between drift velocity and E/P for electrons. w first increases with E/P and then changes very little as E/P increases further.

In this way a multiplication of ions takes place. The electrons are mainly responsible for multiplication. The applied potential at which multiplication just starts marks the end of the ionization region and the beginning of a new region called the proportional region (region 3 in Fig. 3.2).

The number of ions (of one sign) after multiplication can be written as nA, where A is the multiplication factor. In the proportional region A varies between 1 and 10^4. The upper limit of A depends on the number of primary ions. This limit is usually reached when $nA = 10^{11}$. Within these limits, A is independent of the number of primary ions and increases with increasing voltage. In other words, the total number of ions collected is proportional to the number of primary ions, that is, it is proportional to the energy expended by the radiation in the counter.

When the voltage is further increased (region 4 in Fig. 3.2) A increases, but A now depends on n. The increase in A for radiation producing large numbers of primary ions will be less rapid than the increase of A for radiation producing smaller numbers of primary ions. To illustrate this behavior, a second curve is plotted in Fig. 3.2 for a radiation with 2 MeV energy. The numbers of ions collected in the ionization and proportional regions are both proportional to the energy of the radiation. However, beyond the proportional region this proportionality does not hold. Consider the ratio given by

$$\frac{\text{number of ions collected for 2 MeV radiation}}{\text{number of ions collected for 1 MeV radiation}} \qquad (3.1)$$

The ratio equals 2 in the ionization and proportional regions. Beyond these regions the ratio eventually decreases from 2 to 1. The region where this ratio decreases from 2 to 1 is called the region of limited proportionality. There is no direct proportionality between the number of ions collected and the number of primary ions. The number of ions collected does depend, in a more complicated manner, on the number of primary ions. The number of ions collected will be larger for radiation expending higher energy in the counter than for radiation expending less energy.

As the voltage is increased beyond this region, the number of ions collected becomes independent of the number of primary ions. The ratio defined in Eq. (3.1) remains unity in this region. The amplification A increases with voltage, but for any voltage, nA is constant for radiation of any energy. This region (region 5) is designated the Geiger range.

When the ions are collected by the electrodes, the voltage between the electrodes decreases by an increment ΔV from its value before collection according to the relation

$$\Delta V = nAe/C \qquad (3.2)$$

where C is the capacitance of the counter. If an oscilloscope is connected to the electrodes, one will observe a pulse as shown in Fig. 3.3. The pulse has two distinct parts: (1) The first part, depicted in Fig. 3.3, illustrates the rate of rise of the pulse and depends on the speed of the ions and the region in the counter where the ions are produced. (2) The second part shows the rate of decay of the pulse. This rate depends on the time constant RC of the system.

Equation (3.2) shows that the pulse height produced by radiation is proportional to the total number of ions collected nA. Note that this applies to any region. For the ionization and

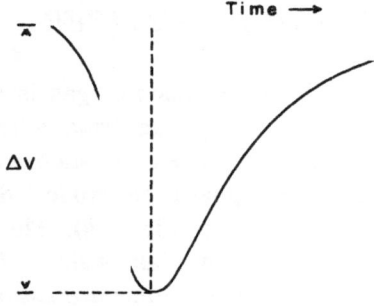

Figure 3-3. Typical change in voltage at the central electrode following the passage of a radiation through the counter.

proportional regions, the pulse height is therefore directly proportional to the energy of the radiation.

3.2. IONIZATION COUNTERS

Ionization counters are used as integrating-type counters or as pulse counters. As we have seen, the rate of discharging of the electrodes is determined by the time constant of the system. If the time constant is made much larger than the collection time of the ions, discharging will be slow compared to the collection time. In this case the potential drop measured at any time will be essentially proportional to the ions collected to that time; in other words, it will be proportional to the energy absorbed by the counter during that time. A counter operating under this condition is an integrating-type counter.

If the time constant is made only slightly larger than the collection time, each radiation will produce a pulse, that is, a voltage change of short duration. In this mode the counter is a pulse counter.

We shall now discuss the three types of ionization counter and their applicability.

3.3. AIR-WALL IONIZATION CHAMBER

An ionization chamber with air as the counter gas is a suitable instrument for the determination of dosage from a radioactive source. A typical ionization chamber used for such a measurement is shown in Fig. 3.4. The guard electrodes define the volume of the counter (shaded area in Fig. 3.4). The sensitive volume is assumed to be bounded by "air walls." Losses of corpuscular radiation from the sensitive volume are assumed to be compensated by the ionization produced in the air walls.

The dosage rate is normally given in units of roentgens. The roentgen (abbreviated r) is defined as an exposure dose of x or gamma radiation such that the associated corpuscular emission

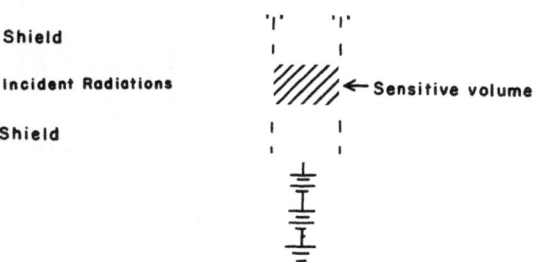

Figure 3-4. Air-wall ionization chamber.

per 0.001293 g of air (this is the mass of 1 cm³ of air at standard conditions) produces, in air, ions carrying 1 esu of quantity of electricity of either sign. From this definition, the dose D (in r), expressed in terms of the volume under standard conditions, can be calculated from the charge collected using the following equation:

$$D = \frac{Q \times 3 \times 10^9 \times T \times 760}{V \times 273 \times P}$$

where Q is the charge collected (in C), T is the temperature (°K), P is the pressure (Torr), and V is the effective volume of the chamber (cm³). The dosage rate R in r/hr can be obtained by measuring the current flow. Since $i = dQ/dt$, the equation for dosage rate is given by (units as above; i in A)

$$R = \frac{i \times 3 \times 10^9 \times 3600 \times T \times 760}{V\ 273 \times P}$$

In this counter, there is no wall to obstruct the radiation and hence, since the volume of air is known, we obtain directly the total amount of ionization produced in a well-defined volume of air.

3.4. POCKET DOSIMETER

A widely used form of ionization chamber is the pocket dosimeter. It consists of an ionization chamber with a self-contained

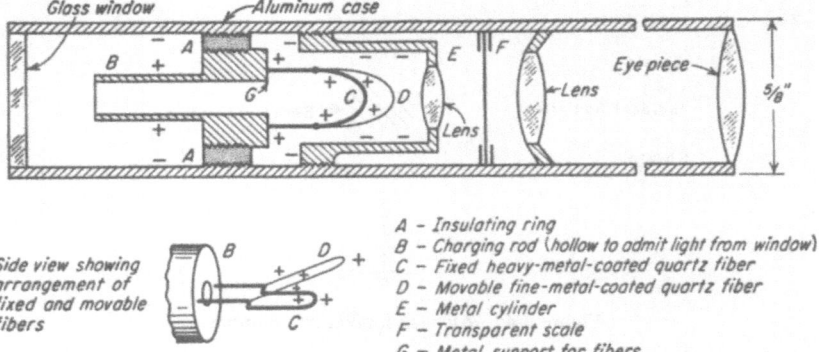

Figure 3-5. Pocket dosimeter. [From: Ralph E. Lapp and Howard L. Andrews, *Nuclear Radiation Physics*, p. 199, Prentice Hall, Englewood Cliffs, New Jersey, 1948.]

charge-reading mechanism. Figure 3.5 shows one form of such a dosimeter. Two metal-coated fibers, one fixed and the other movable, are attached to a cylindrical electrode. The other electrode is the case of the dosimeter. The two electrodes are separated by a nonconductor to minimize charge leakage. The electrodes are charged to the maximum by means of a voltage source until the deflection reaches a position corresponding to the zero reading on the scale. When radiation passes through the dosimeter, the ions formed are collected by the electrodes, decreasing the net charge on them. This decreases the deflection of the fiber and takes it to a position of higher reading on the scale.

3.5. *PULSE-TYPE IONIZATION COUNTER*

Pulse-type counters operating in the ionization region can be used in ionizing particle spectroscopy. They give very good energy resolution (0.7%) for 5-MeV alpha particles. However, a simple two-electrode ionization chamber cannot be used for this purpose.

Assume that a specific type of radiation of 3 MeV is stopped in the sensitive volume of the ionization chamber. It produces approximately 10^5 ions (3×10^6 eV/30 eV). All the radiations of 3 MeV energy will not produce the same number of ions. There will be statistical variations in this number. If one plots the number of ions produced by each radiation versus the number of incident radiations, one gets a plot similar to that given in Fig. 3.6. For the bell-shaped curve shown in this figure, the maximum number of ions produced is 10^5. The spread in the number of ions is usually denoted in terms of the width of the distribution at half-maximum. Since the pulse height is proportional to the number of ions, the pulse height will have a distribution similar to the one in Fig. 3.6. If the distribution is assumed to be Gaussian, the half-width, usually denoted by FWHM (full width at half-maximum), is equal to the standard deviation multiplied by 2.35. Therefore the equation for resolution, R, in percent [defined in Eq. (1.1)]

$$R = (2.35N^{\frac{1}{2}}/N)100$$

gives, for a 3-MeV radiation,

$$R = [2.35 \times (10^5)^{\frac{1}{2}}/10^5]100 = 0.7\%$$

The equation, half-width $= 2.35N^{\frac{1}{2}}$, holds if the ionizations are completely unrelated. However, Fano[1] has shown that the FWHM will be less by a factor of 2–3 than that given by the

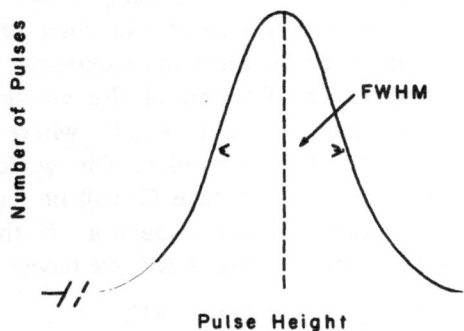

Figure 3-6. Typical pulse height distribution from an ionization chamber. Pulse height is proportional to the number of ions produced by the incident radiation.

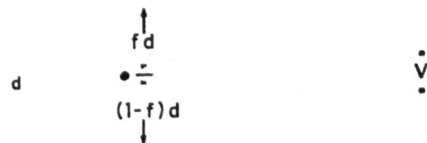

Figure 3-7. An ion pair in a parallel-plate
ionization chamber.

above expression. The half-width is therefore given by the equation

$$\text{FWHM} = F^{\frac{1}{2}}2.35N^{\frac{1}{2}}$$

where F is known as the Fano factor. The Fano factor is estimated to be 0.4 for gas counters and is less than 0.2 for semiconductor counters. Therefore, for the above example, $R =$ 0.7 × 0.64 = 0.5% for gas counters. There are other factors which contribute to a higher value for R. They are (1) noise in the chamber and amplifier, (2) source thickness, (3) variation in rise time of the pulse, and (4) variations in the induced effects of positive ions.

   To understand the shape of the pulse in an ionization chamber, let us consider a simple ionization chamber consisting of two electrodes (see Fig. 3.7). When the ions start to move to the electrodes, the potential of the electrodes changes due to induced charges. Consider an electron produced at a distance fd from this anode, where d is the distance between the electrodes and f is a fraction between 0 and 1. The fraction of the voltage through which the electron will have to travel is fV, where V is the potential difference between the electrodes. The work done on the electron in falling through the voltage fV will be equal to the change in the energy stored on the capacitor. If the voltage change in the collection of the electron is ΔV, we have

$$efV = \tfrac{1}{2}CV^2 - \tfrac{1}{2}C(V - \Delta V)^2$$

Since $V \gg \Delta V$, the term with $(\Delta V)^2$ can be neglected, and we have

$$efV = CV \Delta V \quad \text{or} \quad \Delta V = \frac{fe}{C} \quad (3.3)$$

Equation (3.3) means that an induced charge of fe is collected by the anode. Similarly, a charge $(1 - f)e$ will be induced by the motion of the positive ion to the cathode. The total voltage change upon collection of both charges is

$$\Delta V = \frac{fe}{C} + \frac{(1 - f)e}{C} = \frac{e}{C} \quad (3.4)$$

The way ΔV changes between the formation of the ions and their collection depends on how fast they are traveling. The motion of a charged particle is influenced by the size and mass of the particle, the pressure, the electric field strength, and the chemical composition of the gas. The electron mobility is two to three orders higher than the mobility of positive ions for most of the gases used in ionization chambers. The pulse therefore will have two sections, a fast-rising section due to the electrons and a second section due to the slow positive ions. Figure 3.8 shows a typical voltage signal induced by an ion pair in a plane parallel-plate ionization chamber. (In this case, the time constant is assumed to be infinite.) The three different curves are for three different locations of the ions between the electrodes. However,

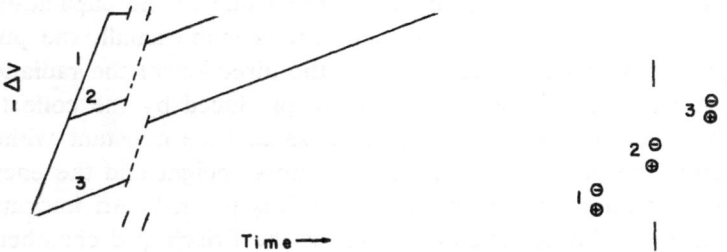

Figure 3-8. Voltage signals induced by ion pairs located at three different positions in an ionization chamber.

Time

Figure 3-9. Voltage pulse due to a radiation
entering an ionization chamber. Infinite time
constant is assumed.

the pulses due to a radiation entering the ionization chamber will look like the one shown in Fig. 3.9. In general, we will not see the sharp breaks in the curves, because of the random changes in the velocity of the ions and the inclination of the track.

In an actual experimental setup, there will be a resistance in parallel with the capacitance of the counter. The height of the output pulse will be proportional to the energy of the radiation only if the time constant is larger than the collection time of the positive ions, which is of the order of 10 msec. Making the time constant large will introduce the following difficulties: (1) Since the pulses are wide, count rate should be kept very low, otherwise "pileup" of pulses will take place. (2) Low-frequency mechanical vibrations of the chamber, which will produce a voltage change due to a change in the value of the capacitance, will be amplified. If the time constant is made small, the pulse height is also found to change with the direction of the radiation.

A chamber where the pulses are produced by the collection of the electrons only can possess a small time constant without affecting the proportionality between pulse height and the energy of the radiation (or the number of primary ions). An ionization chamber suitable for such operation is the Frisch grid chamber, a schematic diagram of which is shown in Fig. 3.10. It has a grid between the two electrodes at a suitable intermediate potential.

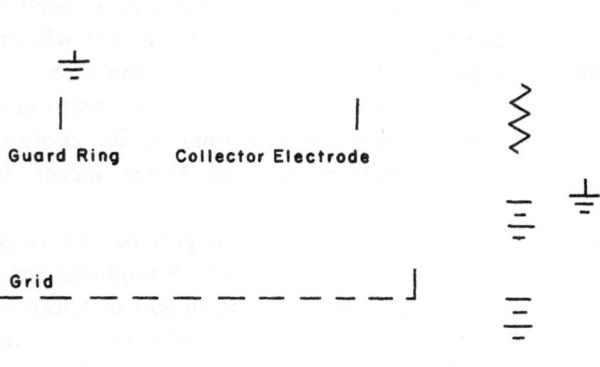

Figure 3-10. Schematic diagram of a Frisch chamber.

The source is kept on the cathode. The pressure of the gas in the chamber and the cathode-to-grid distance are so arranged that all the ionizing radiations are stopped between the cathode and the grid. The positive ions are collected by the cathode; they do not make any contribution to the pulse, since the collecting electrode

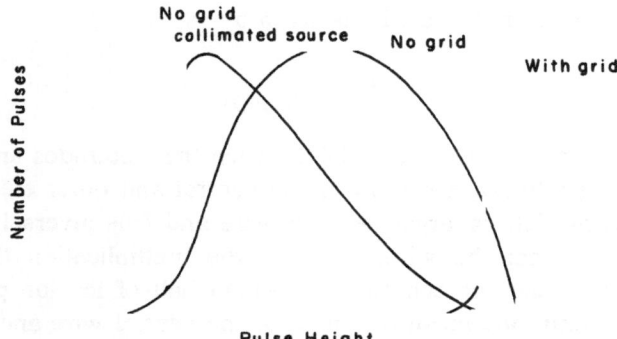

Figure 3-11. The effect of the grid of a Frisch chamber on the pulse height distribution. [From: O. Bunemann, T. E. Crenshaw, and J. A. Harvey, *Can. J. Res. A* **27**, 191 (1949).]

is shielded by the grid. The electrons pass through the grid because of the higher potential at the collecting electrode. The output pulse is entirely due to the electrons, and it will have only the rapidly rising part slightly dependent on the direction of the track. However, since the collection time is very short (1–2 μsec), the output pulse with height proportional to its energy can be obtained for an *RC* constant several times larger than the collecting time.

Figure 3.11 shows the effect of the grid on the pulse height distribution. These results are due to Bunemann *et al.*[2] A chamber without the grid shows a pulse height distribution with a width at half-height of 1.02 MeV. Collimation of the alpha particles improves the resolution by a factor of two, while the gridded chamber shows a width at half-height of 50 keV, which is a great improvement.

3.6. *PROPORTIONAL COUNTER*

A counter working in the proportional region, where the total number of ions produced is proportional to the energy of the radiation, is called a proportional counter. Proportional counters are built with a cylindrical electrode and a central wire, usually tungsten. Because of the geometry of the system, the electric field at a distance x from the wire is given by

$$E = \frac{V}{x \ln(b/a)}$$

where V is the applied potential between the electrodes and a and b are, respectively, the radii of the central and outer electrodes. The electric field is larger near the wire and falls inversely as the distance x from the wire. Most of the multiplication therefore takes place near the central wire. About half of the ion pairs are formed within one mean free path of the central wire and 99% of the ion pairs are formed within seven mean free paths. The collection time of the electrons is very small; however, since the electrons are produced close to the central electrode, ΔV at the central electrode due to the collection of these electrons is very small. Therefore the major contribution to the voltage drop is due to the positive ions. Even though the positive ions are slower than the electrons, they travel through the short distance from the central wire where most of the potential drop occurs in a very short time. The pulse due to an ion pair consequently rises very fast and then slows down. Sometimes there will be an uncertainty in the starting time of these individual pulses if the radial positions of the individual ions relative to the central electrode are different. In such a case, the times taken by different electrons to reach the multiplication region will not be the same. First-stage amplifers integrate the ions to reduce this uncertainty.

If complete proportionality between pulse height and energy is needed, the time constant of the counter amplifer should be at least as great as the collection time of the positive ions, which is about 100 μsec. If one is only interested in detecting the individual particles, much smaller time constants can be used.

In the proportional counter, the ionization is limited to a region surrounding the path of the radiation. Let us assume that radiation 1 enters the counter at time t_1 and that a second, similar radiation 2 enters a different region of the counter at t_2. The potential drop at the collecting electrode will look like the one in Fig. 3.12. If the amplifer scaler can distinguish the voltage change as two pulses and if this is the minimum time separation for which

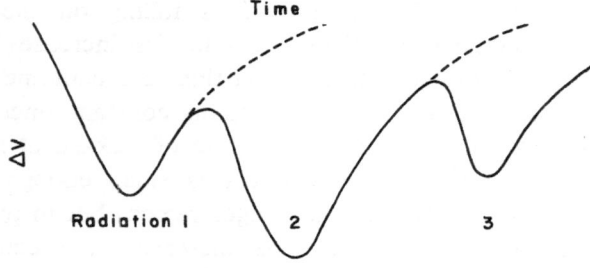

Figure 3-12. Potential drop induced at the collecting electrode by radiations entering the proportional counter at different times. Note the overlap of pulses.

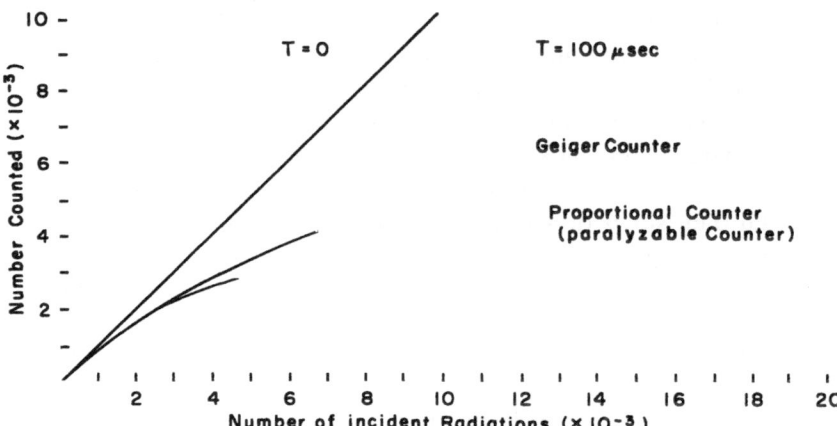

Figure 3-13.   Plot showing the number of counts versus the number of incident
radiations (paralyzable counters are defined in Section 6-3).

this is possible, $t_2 - t_1$ is the resolving time T for the proportional
counter. The resolving time T therefore is dependent on the
electrical system. Equations to calculate the resolving time losses
will be discussed in a later section. It is interesting to see how the
number of counts changes as the number of radiations going into
proportional counter changes. If the resolving time is zero, the
number of counts versus the number of radiations should be a
straight line. However, if the resolving time is finite, this plot will
bend toward the x axis and eventually intersect it (Fig. 3.13).
That is, as the number of radiations falling on the counter
increases, the number of registered counts first increases and after
reaching a maximum goes to zero. At this zero counting rate, the
voltage at the collector electrode remains constant since the rate
of collection of ions is equal to the rate of leakage of ions. The
above discussion is valid only if nA is small enough that the
counter does not operate in the Geiger region, but in practice as
the number of incident radiations increases, the counter will
behave as a Geiger counter. Therefore the count rate will
approach a constant value (see the discussion of resolving time in
Section 3.8).

As noted earlier, the amplification factor of ions A varies from 10^0 to 10^4 in the proportional region. Variation of the amplification for argon with the tube voltage is shown in Fig. 3.14. As can be seen, the increase in A is very fast, especially for low-pressure gases. If an extremely stable power supply is not used, fast amplification changes may produce an increase in FWHM for counters filled with such gases. It is found that addition of polyatomic gases such as CO_2 or CH_4 reduces the fast increase in amplification with voltage. An example is "p-10" gas, which is 10% methane and 90% argon. However, addition of polyatomic gases increases the operating voltage of the counter.

Polyatomic gases limit the life of the counter. The polyatomic

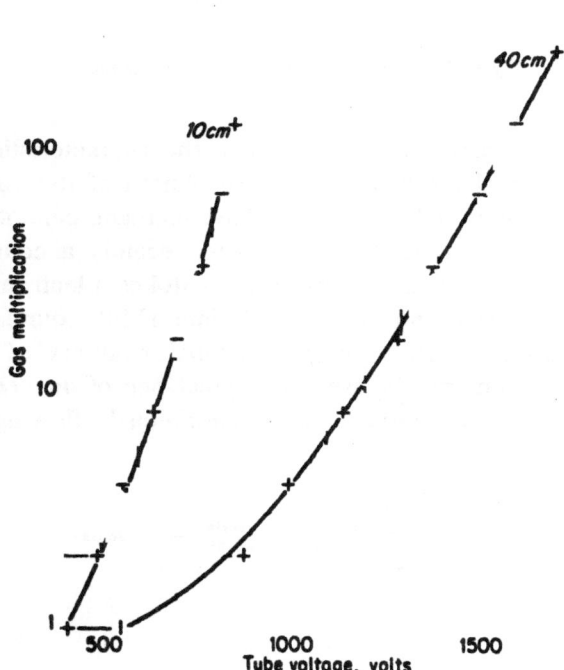

Figure 3-14. Gas multiplication versus voltage for pressures of (a) 100 and (b) 400 Torr; argon 99.6% pure; collector diameter, 0.01 in.; cathode diameter, 0.87 in. [From: B. B. Rossi and H. H. Staub, *Ionization Chambers and Counters*, McGraw-Hill, New York, 1949.]

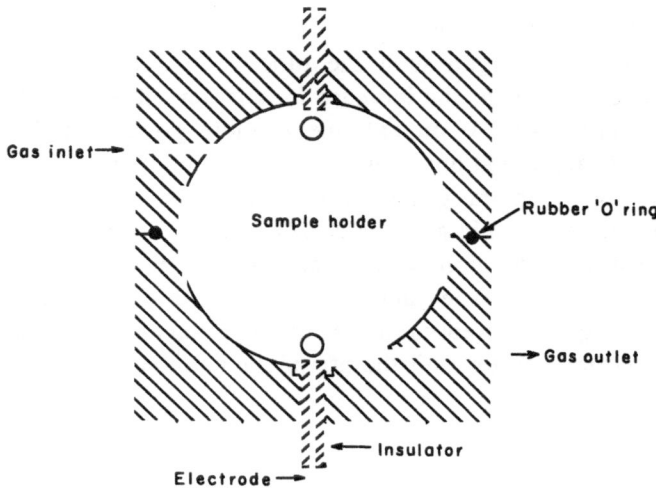

Figure 3-15.   Continuous gas flow 4π counter.

gas molecules break down and change the characteristics of the
tube. This also changes the energy resolution of the counter. A
counter filled with 90% xenon and 10% methane cannot be used
for total counts beyond 10^{12}. However, recently a counter with
95% Xe–3% CO_2 developed by Reuter-Stokes Electronics Components, Inc., was found to have a lifetime of 10^{16} counts.

   ''Windowless'' gas flow proportional counters of different
geometries are commonly used. A typical one of 4π geometry is
shown in Fig. 3.15. Counter gas is continuously flowing through

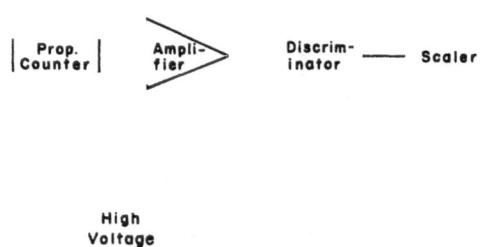

Figure 3-16.   A schematic diagram of a counting
arrangement for a proportional counter.

the counter and the sample can be kept inside the chanber. Some counters have provisions to change the sample without disturbing the gas flow. The counting setup is shown in Fig. 3.16. An amplifier is used with the counter. A typical characteristic curve of counts versus voltage for a proportional counter system using a single source emitting alpha and beta radiation is shown in Fig. 3.17. Pulse heights produced by alpha particles are greater than those produced by beta particles and therefore only alpha particles will be counted at lower voltages. As the voltage is increased, the beta pulses become of sufficient height to be counted along with the alpha particles. Therefore we have a voltage region (first plateau) where only alpha particles are counted and a second region where both alpha and beta particles are counted.

Gas flow counters with thin windows are also in common use in the determination of the energy of alpha and beta particles. They have the advantage that the sample can be introduced into the counter in gaseous form. This is especially useful for the detection of low-energy beta particles emitted from ^{3}H, ^{14}C, etc.

Another point to consider is the relationship between the pulse height and the nature of the particle. The height of the pulses produced by heavily ionizing particles such as alpha particles may differ considerably from that of pulses produced by electrons of the same energy. This difference depends on the

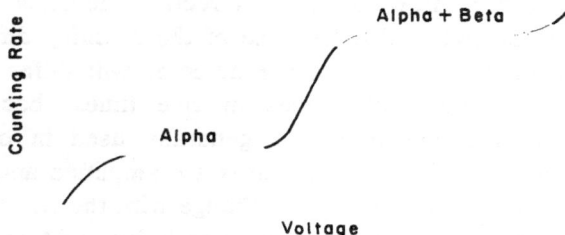

Figure 3-17. Counting rate as a function of operating voltage for the windowless gas flow counter. Two plateaus are observed, one where alpha particles are counted and the second for alpha and beta particles.

nature of the counter, being generally small in gas counters, both proportional and ionization, and semiconductor counters. The energy per ion pair for alpha particles is slightly different from that for beta particles for all the commonly used counter gases. Beta particles have to expend 42.3 eV and alpha particles 42.7 eV to produce an ion pair in helium; the corresponding values are 27.3 eV and 29.2 eV in CH_4. However, this difference is much higher in scintillation counters, and is discussed in the following chapter.

3.7. *POSITION-SENSITIVE PROPORTIONAL COUNTERS*

As pointed out earlier, one of the main differences between the proportional counter and the Geiger counter is that in the proportional counter, ionization is limited to a small region around the path of the incident particle, while in the Geiger counter it spreads throughout the counter. Therefore it is possible to get some information about the position of the incident radiation in the proportional counter. Figure 3.18 is a block diagram of a position-sensitive proportional counter system. The anode is a high-resistance wire (usually a quartz fiber coated with carbon). Assume the particle incident at position X produces ions in the vicinity of the anode. These ions are collected on the anode and cause current flows in both directions through the anode. The amount of current flow in each direction depends on the resistances of the two paths. Because of the resulting difference, pulses produced at the ends of the detector will differ in their height and rise time. Differences in rise times, because of differences in time constants, are generally used in obtaining information about positions. The pulses are amplified and doubly differentiated. The crossover points change with the rise time and therefore the difference in the crossover points will reflect the change in the resistance along the two paths of the anode. The time-to-pulse amplitude converter produces a pulse with an amplitude dependent upon the time difference between the pulses

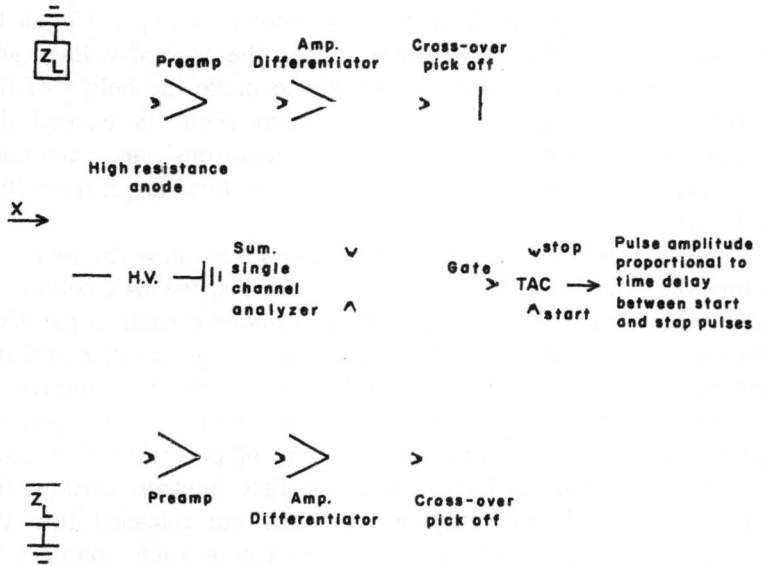

Figure 3-18. A block diagram of a position-sensitive proportional counter.

going into it, and therefore the height of the pulse will be a measure of the position of the incident radiation. Position resolution of the order of 0.1 cm has been obtained for a 20-cm-long counter.[3]

3.8. *NEUTRON COUNTERS*

In addition to using proportional counters for charged particles like alpha and beta particles, they can be used to detect neutrons. A typical neutron counter usually contains pure BF_3 or a mixture of BF_3 with any of the standard counter gases. When a thermal neutron is absorbed by the boron nucleus, two heavily ionizing particles, an alpha particle and a recoiling lithium nucleus, are released. The pulses produced by the reaction products are high compared to the pulses produced by background radiations such as gamma rays.

A counting system similar to that shown in Fig. 3.16 can be used for the detection of neutrons. When the applied voltage and the amplification are adjusted properly to make the height of the pulses produced by the neutron reaction products exceed the discriminator level, essentially only neutrons are counted. Counters of this type can be used for a wide flux range, from 10^{-3} to 10^5 n/cm²-sec.

One method of detecting fast neutrons is to slow the neutrons before they reach the BF_3 counter. To do this, the BF_3 counter is usually covered with some hydrogenous material such as paraffin. The elastic collisions between the protons in the paraffin and the fast neutrons slow the neutrons before they enter the counter. A second type of counter for detecting fast neutrons is the proton-recoil counter. This counter has a lining of polyethylene around the inside of the cathode. When a fast neutron strikes the hydrogen atom in polyethylene, protons are released into the counter by elastic collisions. The response of such counters to neutrons of various energies depends very strongly on the construction. It is also possible to design counters whose response to neutrons of various energies very closely approximates the response of biological tissue; such tubes are used in survey instruments.

In counters used for the detection of thermal neutrons the counter is operated as an ionization chamber and contains fissionable material. When thermal neutrons are absorbed by fissile nuclei, they produce fission products having energy on the average of ~50 MeV each. The pulses produced are therefore easy to detect and can be readily distinguished from the low-energy background radiations.

3.9. GEIGER–MÜLLER COUNTERS

A widely used gas counter is the Geiger–Müller (G–M) counter, also known as the Geiger counter. It has several advantages, such as high efficiency for beta particles and output pulses of larger pulse height. In addition, no amplifying system is needed, thereby reducing the cost of the counting system. A disadvantage lies in the fact that output pulses are of the same height and therefore no information about the energy of the radiation can be obtained from these counters. Figure 3.19 shows the details of construction of a G–M counter.

In the Geiger counter, the negative electrons travel toward the central electrode and the multiplication, sometimes called the Townsend avalanche, takes place over a short distance near the anode. The electrons are collected by the anode in a few microseconds. The photons produced in the deexcitation of the atoms spread the ionization along the central wire in the counter. This is one major difference between the proportional counter and the Geiger counter. The spreading of the ionization through the

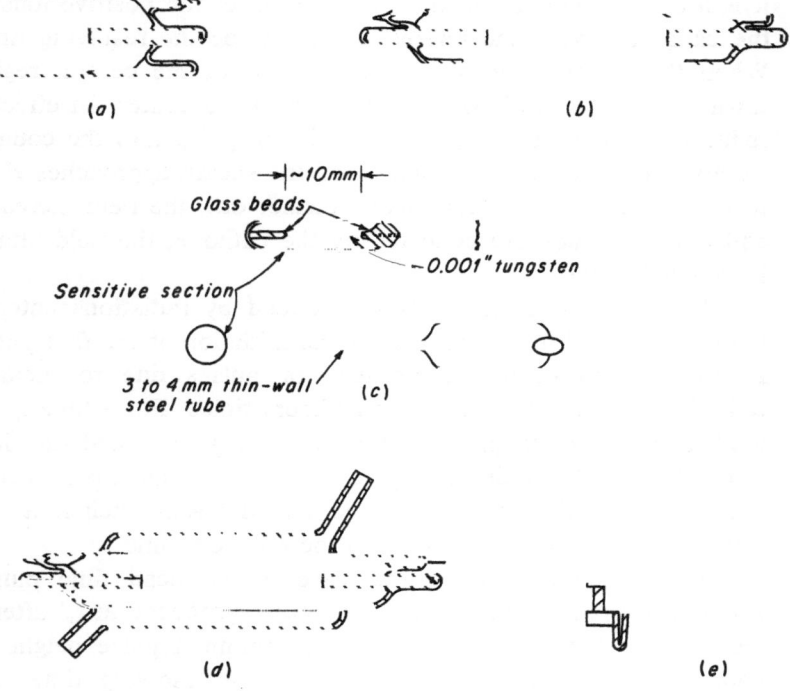

Figure 3-19. The details of construction of a typical G–M counter.

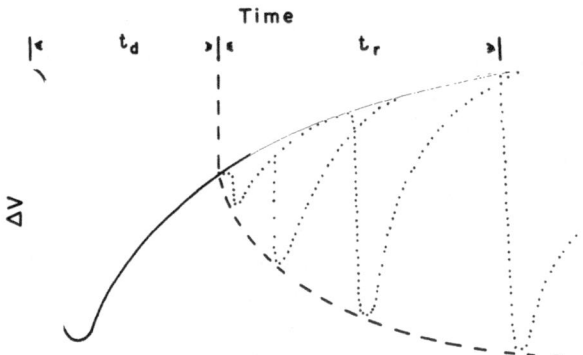

Figure 3-20. Pulse shapes of pulses arising from radiations
entering a Geiger counter at different times.

length of the counter and the slow motion of the positive ions to
the cathode have some interesting effects on the resolving time.
When the positive ion sheath moves out of the central region
toward the cathode, it shields the field in the center. In effect it
reduces the field so that another radiation going into the counter
cannot start a second avalanche until the sheath approaches close
to the cathode. As the ions move farther out, the field increases
and finally as they are collected by the cathode, the field attains
its original value.

Let us consider the pulses produced by radiations entering
the counter at different times after the initiation of the first pulse.
Figure 3.20 illustrates the shapes of pulses due to incident
radiation entering the counter at different times. For a time t_d the
field is too weak to produce any pulse. This is called the dead
time. After t_d the radiations produce pulses of increasing height
and at $t_d + t_r$ the pulse produced has the same height as the
original pulse. The time t_r is called the recovery time.

The resolving time in this case is the dead time plus a
fraction of the recovery time. If the pulse produced at t_r' after t_d
has a pulse height equal to ΔV, the minimum pulse height for
counting pulses, then $T = t_d + t_r'$ is the resolving time. The
above discussion assumes that the only changes in potential at the

electrodes are during the collection of charges. It used to be a practice to reduce the potential for a fixed time immediately after a pulse was registered. This was done to reduce the number of secondary pulses. In modern counters this is achieved by adding quenching gases (see Section 3.10).

At this point, it is appropriate to discuss how to make corrections for counting losses due to "coincidence loss." Coincidence loss is the loss in counts due to radiations reaching the counter within time intervals shorter than the resolving time. Let us assume that n radiations per second enter the counter and n' counts are obtained. Since n' is the number of counts for a time $n'T$, the total time during which the detector is counting is $1 - n'T$ and therefore we get the relationship between n and n' as

$$n = n'/(1 - n'T) \tag{3.5}$$

The coincidence loss, that is, the loss in counting due to finite resolving time, is

$$n - n' = (n')^2 T/(1 - n'T) \tag{3.6}$$

In first approximation, counting losses increase in direct proportion to the square of the number of incident radiations and in proportion to the resolving time. A plot of n versus n' is shown in Fig. 3.13. The maximum number of counts one can get from a Geiger counter is given by $n_{max} = 1/T$.

A frequently used method of determining the resolving time is the split source method. In this method, the source 1 and a blank are first counted (Fig. 3.21). The blank is of the same

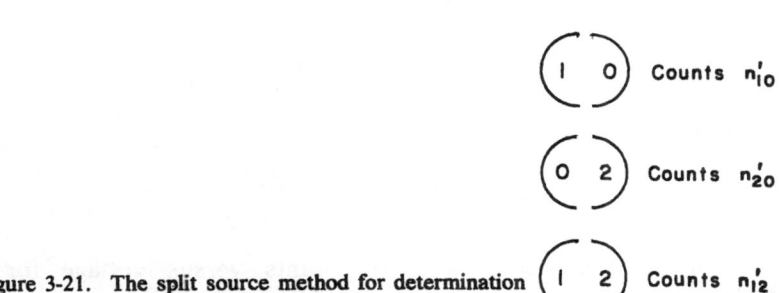

Figure 3-21. The split source method for determination of resolving time.

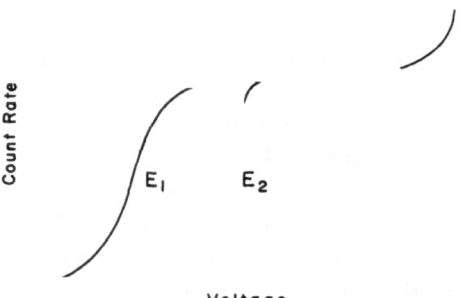

Figure 3-22. Characteristic curve of count rate
versus voltage for a Geiger counter.

material as and is geometrically identical to the source to keep the
scattering, etc., the same throughout the experiment. Then the
second source and the blank and finally the two sources together
are counted. If n_1', n_2', and n_{12}' are the counts obtained,
respectively, and if n_1, n_2, and n_{12} are the number of radiations
entering the counter, respectively, we have the relations

$$n_1 = \frac{n_1'}{1 - n_1'T}$$

$$n_2 = \frac{n_2}{1 - n_2°T}$$

$$n_{12} = \frac{n_{12}'}{1 - n_{12}°T}$$

Since $n_1 + n_2 = n_{12}$, we have

$$\frac{n_1'}{1 - n_1'T} + \frac{n_2'}{1 - n_2'T} = \frac{n_{12}'}{1 - n_{12}'T}$$

From this equation, we get

$$\tau = \frac{n_1' + n_2' - n_{12}'}{(n_{12}')^2 - (n_1')^2 - (n_2')^2}$$

The characteristic curve of counts versus voltage for a
Geiger counter gives much information about the counter. The

characteristic curve can be obtained by keeping a radioactive source of long half-life near the counter and taking the counts for a fixed time for various applied voltages on the counter. Such a curve for a typical counter is shown in Fig. 3.22. The scaler, which registers the counts, determines the nature of the beginning part of the curve. The scaler needs a pulse of a certain minimum height for it to register. Let us consider the case where this minimum pulse height is E_1 volts. The position of E_1 with respect to the pulse height produced by a counter in the limited proportionality and Geiger regions is shown in Fig. 3.23. If curve 1 shows the pulse height for different voltages for the highest energy radiation from a source, the counter will start counting at voltage V_1 when curve 1 crosses E_1, in other words, when the pulses produced by these radiations have a pulse height above that required for counting. As the voltage increases, more and more radiations will be counted, and eventually at voltage V_1 , for which all the pulses have height above E_1, all the radiations will be counted. The part of the characteristic curve between the voltages V_1 and V_1' depends on the energy distribution of the radiation reaching the counter. Above V_1' all the pulses will be counted and therefore the number of counts remains constant. The region where the number of counts does not change with applied voltage is called the plateau of the characteristic curve. If the minimum pulse height required to be counted is E_2, the characteristic curve, as depicted in Fig. 3.23, will be rapidly

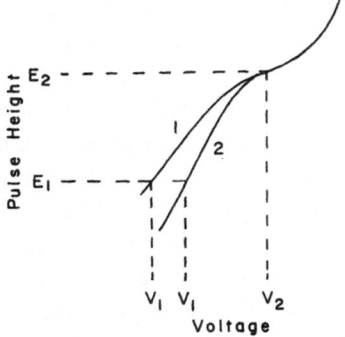

Figure 3-23. Pulse height versus voltage of a counter in the limited proportionality and Geiger regions. Curve 1 is for the radiation with the highest energy from the source and 2 is for the radiation with the lowest energy.

rising. The starting potential in this case is V_2. If the scaler is sensitive for smaller pulses, the plateau will be longer and the characteristic curve will have a slowly rising part. At the end of the plateau, we get a sudden increase in the number of counts. Here the counter is operating in the discharge region.

3.10. COUNTER GAS

Any gas can be used in a counter. However, better performance results if the gas satisfies the following requirements: (1) The operating potential should not be too large; (2) the gas should not form negative ions; and (3) the gas should not have metastable states. Some gases, such as chlorine and air, have a high affinity for electrons and form negative ions easily. Such gases are not good for use in counters. The velocity of a negative ion is approximately equal to the velocity of the positive ion. If a negative ion is formed not very close to the central electrode, it will reach the high-field region at a time when the positive ions have reached the outer electrode. The field in the counter will have regained enough to start a second avalanche. Thus the negative ion will produce a pulse, which will be a satellite pulse with respect to the pulse produced by the incident radiation. This is called a spurious pulse.

Satellite pulses can also arise from the existence of metastable states. Such states are atomic excited states with long lifetimes. They are formed during collisions with high-energy electrons. Deexcitation of these states results in the emission of photons. The photons so produced could release an electron by the photoelectric effect. If this happens after the collection of the positive ions, a satellite pulse will be produced.

There is another mechanism by which satellite or multiple pulses can be produced. When positive ions are collected by the cathode, they become neutralized by picking up electrons from the cathode, and energy equivalent to the ionization energy will be available. This energy could result in a photon or the release of an electron from the metal by secondary emission if the energy is higher than the work function of the metal. This electron could start a second pulse. The photon could also emit a photoelectron, which, in turn, could initiate a second pulse. Whatever the process of energy transfer, there is a chance of initiating a second pulse when the positive ions are collected by the electrode. This process can repeat itself indefinitely.

The production of a secondary pulse by a metastable atom or during neutralization of a positive ion can be controlled by adding "quenching" gas to the counter gas. Quenching gases are polyatomic gases, common ones being ethyl alcohol vapor, ethyl formate, and halogen gases. To understand the working of a quenching gas, let us assume we have a counter with an Ar–alcohol mixture in it. In a collision of an argon ion with the alcohol molecule, an electron will transfer to the argon ion from the alcohol molecule since the ionization potential of argon (15.7 eV) is higher than the ionization potential of the alcohol (11.3 eV). The excess energy is used up in breaking up the alcohol molecule. An alcohol ion reaching the electrode picks up an electron, with the release of energy. This energy also dissociates the alcohol molecule and hence, once again, there is no second avalanche. Similarly, when an alcohol molecule collides with an argon atom in its metastable state, the excitation energy of argon is transferred to the alcohol molecule. In this case also the alcohol molecule breaks up and again no photon or electron is released.

Since the alcohol molecule stops the release of photons or electrons, the chance of satellite pulses occurring is reduced. However, as the quenching gas is used up, the number of satellite pulses increases and eventually the counter becomes unusable. A counter in which the quenching gas has been used up to a large extent will have a plateau with a higher slope. Therefore a test for a good counter is the flatness of the plateau.

Organic gases such as alcohol, and halogen gases such as chlorine, are common quenching gases. Halogen-quenched tubes have longer lifetimes. The diatomic halogen molecules dissociate during quenching but recombination takes place, replenishing the supply of quenching gas. However, the plateau of a halogen-quenched counter is short, with a rather high slope.

Table 3.1. Absorption of Beta Rays of Different End-Point Energies
E_{max} in the Counter Window

		Absorption in the window, %		
Isotope	E_{max}, MeV	30 mg/cm²	20 mg/cm²	3 mg/cm²
^{11}C	0.95	50.8	37.6	7
^{14}C	0.15	99.99	99.76	60
^{18}F	0.7	65.4	50.8	11
^{32}P	1.69	28	19.7	4
^{42}K	3.5	11.7	8	2

3.11. DETECTION OF ALPHA AND BETA RADIATIONS BY GEIGER COUNTERS

The Geiger counter is capable of giving an output pulse even if only one primary ion pair is produced. Therefore if the radiation can enter the sensitive volume of the counter, it will be counted. Hence Geiger counters have 100% efficiency for ionizing radiations of energy down to 30 eV. However, this high efficiency can be achieved only if the source is introduced into the counter. Even in such experimental arrangements corrections have to be made for wall effects (sources decaying near the wall). The internal counting method is especially suitable for low-energy beta sources. When the radioactive source is kept outside the counter, special consideration has to be given to the thickness of the counter window. Since the ranges of alpha and beta particles are very small, counters should be provided with very thin windows. The energy of alpha particles from a radioactive source lies between 4 and 10 MeV; hence a window thickness slightly smaller than the range for a 4-MeV alpha particle, 2 mg/cm², will be sufficient for alpha detectors. The energy distribution for beta particles is continuous and therefore, whatever the thickness of the counter window, a fraction of the beta ray will be stopped in the window. This fraction depends on the thickness of the window and also on the radioactive source, since the maximum energy of the beta ray and its energy distribution change from source to source. Table 3.1 shows the fraction of beta ray absorbed for different thicknesses of some commonly used isotopes. A typical beta counter has a window thickness of 1–3 mg/cm². Mica, Mylar, and stainless steel are commonly used for windows.

3.12. *DETECTION OF X RAYS AND GAMMA RAYS BY GAS COUNTERS*

X rays, because of their low energies compared to gamma rays, have a high photoelectric absorption cross section. Hence it is necessary that the gas counter be provided with a thin window. Thin windows of the type used for beta counters can be used for x-ray counters. Sometimes beryllium windows of 1–3 mm thickness are used. Since beryllium is a material of low atomic number, its photoelectric absorption cross section is comparatively small. Therefore a good fraction of the x rays will enter a counter provided with such windows. For x rays up to 20 keV very good detection efficiency can be obtained. Figure 3.24 shows the variation of efficiency for a typical x-ray detector. The efficiency in the low-energy region decreases because of absorption in the window and in the high-energy region because of the decrease in the photoelectric absorption cross section.

The gas counter can also be used for the detection of gamma rays. However, the efficiency of such counters is less than that for

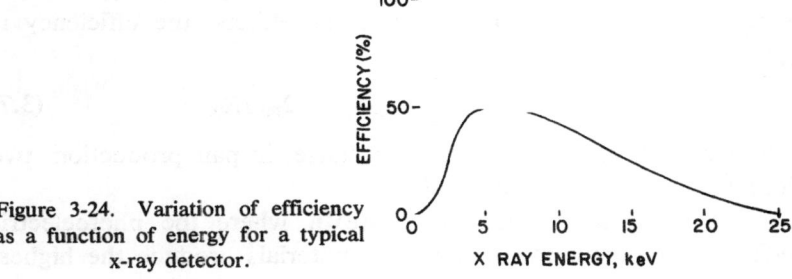

Figure 3-24. Variation of efficiency as a function of energy for a typical x-ray detector.

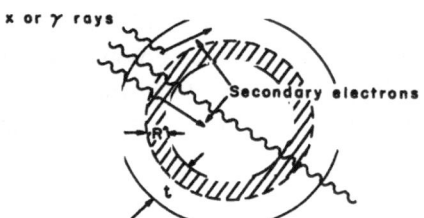

x or γ rays

Secondary electrons

Figure 3-25. Interaction of gamma rays in the wall of the counter. [From: J. W. Price, *Nuclear Radiation Detection*, McGraw-Hill, New York, 1958.]

x rays, for the following reason. As the energy of the gamma rays increases, the cross sections for photoelectric and Compton absorption decrease and the fraction of gamma rays that produce a secondary electron in the counter gas decreases very fast. Therefore we depend on secondary electrons produced in the counter wall to ionize the gas. The calculation of the efficiency ϵ for gamma rays in these circumstances is difficult to make because of the variation in the direction and in the energy of the secondary electrons and their crooked paths in the walls. However, to get an approximate expression, let us consider the gamma rays entering the counter as shown in Fig. 3.25.

Suppose $_{PE}\mu_l$ is the linear absorption cross section for the photoelectric effect and R_{PE} is the range of these electrons. If the thickness is not much more than R_{PE}, the fraction of gamma rays producing photoelectrons is $1 - \exp(-_{PE}\mu_l R_{PE})$. Since the electrons can be emitted in all directions, only half of them will enter the counter. Therefore the fraction of gamma rays producing photoelectrons that go into the counter from one side of the wall is $\frac{1}{2}_{PE}\mu_l R_{PE}$ and the fraction from the opposite side is also $\frac{1}{2}_{PE}\mu_l R_{PE}$. Similarly if $_C\mu$ and $_{PP}\mu$ are absorption coefficients for the Compton effect and pair production and R_C and R_{PP} are the ranges of electrons produced by these effects, the efficiency is given by

$$\epsilon = {}_{PE}\mu_l R_{PE} + {}_C\mu R_C + 2_{PP}\mu R_{PP} \tag{3.7}$$

The last term is multiplied by 2 because, in pair production, two ionizing particles are produced.

At low energies (up to 1 MeV), where the photoelectric effect is predominant, high-Z wall materials produce the highest efficiency counters since the cross section for photoelectric absorption is proportional to Z^5. In the intermediate energy region (1–5 MeV), the Compton effect is predominant and the material of the wall has little effect on the intrinsic efficiency since the Compton absorption coefficient is proportional to Z and the range of the electrons is inversely proportional to Z. At higher energies (above 3 MeV), where pair production predominates, the efficiency increases with Z. The dependence of efficiency on energy for three different cathode materials is shown in Fig. 3.26. The dashed curve is the efficiency calculated for aluminum using Eq. (3.7). Efficiency is very low at low energies, even though the photoelectric cross section is very high. The reason for this is the low range of electrons produced by the low-energy photons. Note that this equation is only an approximation.

3.13. APPLICATION OF INTERNAL COUNTING METHODS

As mentioned in Section 3.11, high detection efficiency can be achieved by internal counting methods. An example of this, the 4π-counting method, is discussed in Section 3.6. This technique

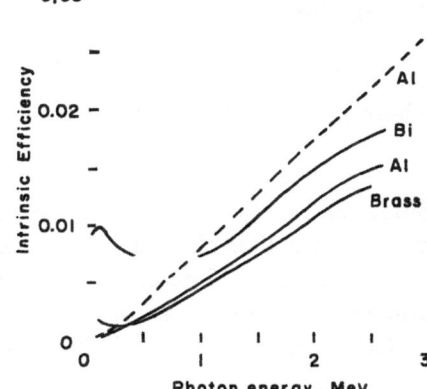

Figure 3-26. Dependence of efficiency on energy for three cathode materials. The dashed curve represents the theoretical efficiency for aluminum. [From: H. Brandt *et al.*, *Helv. Phys. Acta* **19**, 77 (1946).]

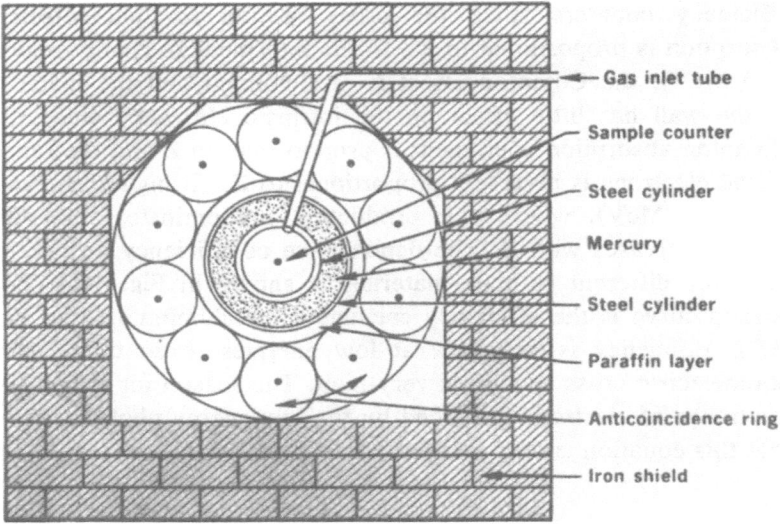

Figure 3-27.   Shielding arrangement for proportional counters used in dating.
[From: Henry Faul, *Nuclear Clocks*, Atomic Energy Commission, Division of
Technical Information 1966; 1968 (Rev.)]

has also improved the sensitivity of radioactive dating, which is
now discussed.

Neutrons in cosmic rays reaching the earth's atmosphere at a
constant rate react with nitrogen atoms and produce ^{14}C according
the reaction

$$^{1}n + {}^{14}N \rightarrow {}^{14}C + p$$

where p is a proton. Carbon-14 decays with a half-life of 5568 yr:

$$^{14}C \rightarrow {}^{14}N + e$$

Since the reaction time, from the beginning of the solar system, is
much longer than the half-life, the activity has the maximum value
[see Eq. (2.22)]. Therefore the specific activity of carbon in the
atmosphere (CO_2) is a constant. Because of continuous exchange
of CO_2 between air and living systems, living systems will have

the same constant specific activity, which, as determined by Libby, is 15.3 ± 0.5 dpm/g. Once the system is dead, activity decreases according to

$$SA = 15.3e^{-0.6936/5568t}$$

where SA is the specific activity in dpm/g carbon, and t is the time since the death of the sample.

Carbon in the sample whose age is to be determined is introduced into the proportional counter after being reduced to a gas. The carbon is either converted into CO_2 by burning it in oxygen or is chemically changed into methane, ethane, or acetylene.

Since the activity is very low, special precautions have to be taken to reduce the background. A properly shielded experimental arrangement is shown in Fig. 3.27. Iron or lead shielding more than 1 ft thick stops much of the cosmic rays and radiation from contaminations in the laboratory. Only neutrons and high-energy radiation go through the shielding. A ring of Geiger counters is placed inside this shield. These counters are connected in anticoincidence with the main counter; if a radiation goes through one of these guarding counters and the main counter, the signal from the main counter will be cancelled electronically by the anticoincidence circuit. Thus radiations from the outside are not counted. Further, shieldings of paraffin and mercury are used around the main counter. Paraffin slows the neutrons and captures them, while highly purified mercury further shields the counter from unwanted radiations. Internal counting in liquid scintillants is also in use.

REFERENCES

1. U. Fano, *Phys. Rev.* **72**, 26 (1947).
2. O. Bunemann, T. E. Crenshaw, and J. A. Harvey, *Can. J. Res. A* **27**, 191 (1949).
3. C. J. Borkowski and M. K. Kopp, *Rev. Sci. Instrum.* **39**, 1515 (1968).

BIBLIOGRAPHY

Serge A. Korff, *Electron and Nuclear Counters*, Van Nostrand, New York, 1955.

W. J. Price, *Nuclear Radiation Detection*, 2nd ed., McGraw-Hill, New York, 1964.

E. K. Ralph and H. N. Michael, Twenty-five years of radiocarbon dating, *Am. Scientist* **62,** 553 (1974).

H. H. Staub, in *Experimental Nuclear Physics* (E. Segre, ed.), Vol. 1, Wiley, New York, 1953.

4

Scintillation Counters

When radiations fall on certain materials called "scintillants," flashes of light are produced. Detecting these flashes with the naked eye or with the help of optical instruments was one of the oldest methods of radiation detection. Rutherford, using a ZnS screen as a scintillant, employed this method to count the scattered alpha particles in his historic alpha-scattering experiment. This method is tedious and very crude and was soon replaced by the use of gas counters, where the counting is done electronically and some information about the energy of the radiation can be obtained if needed. In 1944, Curran and Baker replaced the use of the naked eye by a photomultiplier and later Kallmann replaced the small, thin crystals of ZnS by naphthalene. These two changes revolutionized scintillation detection, making it possible to detect, record, and analyze electronically the pulses produced by individual radiations.

4.1. GENERAL CHARACTERISTICS OF SCINTILLATION COUNTING SYSTEMS

A scintillation counting system consists of a scintillant, a photomultiplier, a power supply, and an amplifier–analyzer–scaler

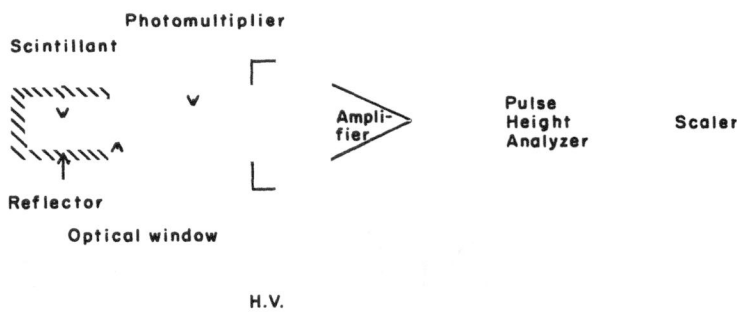

Figure 4-1.   A schematic diagram of a scintillation counting system.

system (Fig. 4.1). When an ionizing radiation passes through the scintillant, it produces photons. The mechanism involved in the production of light is complicated and not well understood. Also, the mechanism is different for different types of scintillants. A discussion of this and other properties of different scintillants is given in a later section. The number of photons produced n_p is proportional to the energy absorbed in the scintillant. The scintillant is covered with a reflector except on the side connected to the photomultiplier, as indicated in Fig. 4.1.

The photomultiplier has a photoelectric film (usually coated onto the photomultiplier tube) as its first element. When light falls on it, electrons are released. The number of electrons produced is proportional to the number of photons falling on the photocathode (which is equal to $n_p\epsilon_c$, where ϵ_c is the collection efficiency of the system) and is equal to $n_p\epsilon_c\epsilon$, where ϵ is the efficiency of the photoelectric material. In most cases ϵ is about 10%. The electrons produced are focused onto the first dynode. The dynodes are coated with a material like cesium antimonide so that, when high-energy electrons hit them, secondary electrons are produced and hence electron multiplication occurs. The photomultipliers have ten or more dynodes. When an electrical potential is applied between any two dynodes, the secondary electrons produced in preceding dynodes gain energy before they strike the next dynode and produce more secondary electrons.

Thus multiplication occurs in each dynode and the final number of electrons N_e collected by the last electrode, called the collector, is given by

$$N_e = n_p \epsilon_c \epsilon m_1 m_2 \cdots \qquad (4.1)$$

where $m_1 m_2, \ldots$ are the multiplication factors for successive dynodes. The factors $m_1, m_2, \ldots$ are dependent on the potential between the dynodes and are usually independent of the number of incident electrons on any dynode. From Eq. (4.1), one can see that N_e is proportional to n_p and, consequently, is proportional to the energy which the incident radiation can supply to the scintillant. The electrons finally pass through a resistance, producing a voltage drop across it. The voltage drop is of short duration and is proportional to N_e. This electrical pulse can be amplified and analyzed.

Scintillation counters can be compared with proportional counters. Both give an output pulse proportional to the energy of the radiation. The multiplication in gas counters is due to the multiplication of ions, while in scintillation counters it is due to the production of secondary electrons at the dynodes. The duration of the pulse is shorter for scintillation counters than for proportional counters.

4.2. *SCINTILLANT MATERIALS AND MECHANISM OF SCINTILLATION*

Some materials can absorb energy and reemit part of the energy in the form of light. This process is called luminescence. Materials reemitting the radiation in a time of the order of microseconds or less are called fluorescent. Materials in which the time delay between absorption and emission is longer are called phosphorescent. In the detection of radiation, only fluorescent materials are used; when applied for this purpose, they are called scintillants. A required feature of the scintillant is that it should be highly transparent to the emitted photons. The fraction of photons that are absorbed by the scintillant varies from

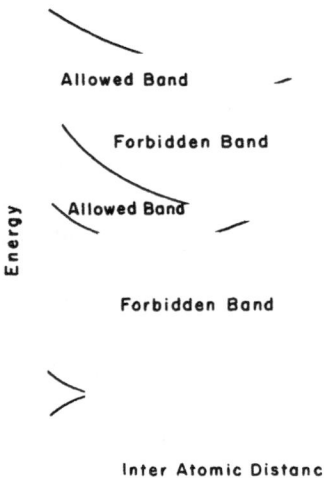

Figure 4-2. Broadening of energy levels as a function of interatomic distance. The band is made up of closely spaced levels.

material to material. Inorganic scintillants are almost 100% transparent. Organic scintillants are in general less transparent.

Several types of scintillants are used: inorganic solids, mainly alkali iodides with or without impurities; organic solids, largely aromatic hydrocarbons, both unsubstituted and substituted; and organic solutions in liquids or in plastic solvents.* The mechanism of scintillation is not completely understood and it is beyond the scope of this book to attempt a detailed picture. The following qualitative discussion may help one to comprehend some of the basic properties.

In a free atom the electrons occupy energy levels defined by certain quantum numbers and having specific value for their energy. When atoms are brought together in solid materials the atomic electrons, especially the outer ones, interact with each other. This interaction drastically changes the energy level diagram. The energy levels now broaden to form bands. The extent of broadening of these bands depends on the extent of interaction.

* Noble gases (in both the gaseous and liquid state) and nitrogen have recently been used as scintillants for special uses. Their properties are not discussed here. (See Ref. 4.)

Hence the broadening is smaller for the inner electrons. Figure 4.2 shows the broadening of the energy levels as a function of interatomic distance. The gap between the bands is called the forbidden band or the band gap. The name means that electrons are not allowed to occupy this gap. However, one will find electronic levels of impurity atoms in the forbidden band. These levels play an important part in determining the electrical and luminescent properties of materials.

Figure 4.3 is a typical energy band diagram. The band filled with electrons is called the valence band, and the empty band is called the conduction band (in metals these bands are half-filled). The electrons in the conduction band are free to move around in a solid. These free electrons, along with the holes (discussed below), are responsible for the electrical conductivity of the material. By supplying sufficient energy, at least equal to the width of the band gap, one can raise an electron to the conduction band. The vacancy thus created in the valence band is called a hole. The hole moves around in the valence band and behaves like a positively charged particle, contributing to the conductivity of the material. Impurity atoms (such as thallium in NaI crystals) give rise to electronic energy levels in the forbidden band (Fig. 4.3), and these levels can be called the activator levels. Some-

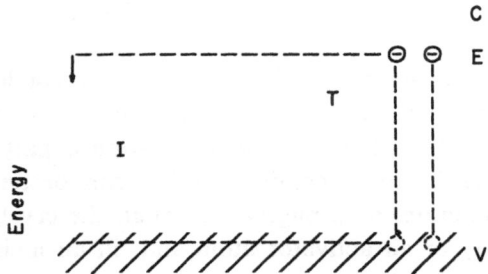

Figure 4-3. Electronic energy band diagram of an ionic crystal showing traps and activator centers. *V* is the valence band; *C* the conduction band; *E* the exciton band; *T* the trap center; and *I* the electronic level of activator atoms.

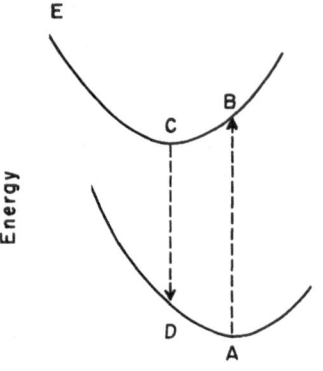

Energy

D
A

Inter Atomic Distance

Figure 4-4. Electronic energy levels of an activator center as a function of the interatomic distance.

times electrons in the valence band can be raised to a level just below the conduction band. The excited electron and the hole form a bound system due to the electrostatic attraction between them. Such bound systems are called excitons. To a good approximation the energy levels of the excitons are hydrogenlike. Even though the bound exciton system can move around in the crystal, it does not contribute to conductivity, because its net charge is zero. Van Sciver[1] has discussed the role of excitons in activated and pure NaI crystals. He attributes the luminescence in pure NaI at 3030 Å to exciton annihilation. There are other levels arising from impurities or defects that act as electron, hole, or exciton traps.

The mechanism by which the energy of a conduction electron or an exciton is converted to a photon will now be considered. As previously mentioned, the incident radiation imparts energy to an electron, resulting in a conduction electron or an exciton. The electron or exciton, as it migrates through the crystal, may arrive in the vicinity of an activator atom and, in an inelastic collision, excite an electron of the atom into one of the excited states. The electron so excited will normally return within 10^{-8} sec to the ground state and thereby emit a photon.

In order to account for the transparency of the scintillant and the temperature dependence of the efficiency of photon production, let us refer to Fig. 4.4. This figure depicts two electronic energy levels of an activator center as a function of the interatomic spacing. Each curve exhibits a minimum at a different interatomic spacing. The system normally has minimum energy and according to the Franck–Condon principle,[2] the emission or absorption of energy will be accompanied by a transition which leaves the interatomic spacing essentially unchanged. Such transitions are represented by the vertical lines, AB representing absorption and CD emission. It is evident that, since the absorption and emission energies are unequal, the frequencies of absorption and emission differ. For this reason, the material will not usually absorb the photon that it emits.

Instead of emitting photons, the scintillant may undergo radiationless transitions. Such a transition will take place if the crystal can supply energy, usually in the form of phonon energy, equal to the energy difference between the points C and E. The probability of this type of transition increases with increasing temperature and is proportional to $e^{-\Delta E/kT}$, where ΔE is the energy difference between points C and E. The number of photons emitted therefore decreases with increasing probability of nonradiative transitions. From consideration of the factor $e^{-\Delta E/kT}$, we see that as the temperature T increases, the output light intensity decreases. This result has been observed for all inorganic scintillants.

Each electronic level has its characteristic decay time T_d and decay constant $\lambda = 1/T_d$. The parameter λ is dependent on temperature and concentration. If all the photons are emitted from the same level, the light output will decay exponentially as given by the equation

$$n = n_p e^{-\lambda t} \qquad (4.2)$$

This equation will be used later to obtain the shape of the pulse produced by scintillation counters.

We shall now consider some of the properties of organic scintillants. Luminescence of organic compounds, especially of the aromatic compounds used in scintillation detectors, is a molecular phenomenon. The π electrons in these compounds (like

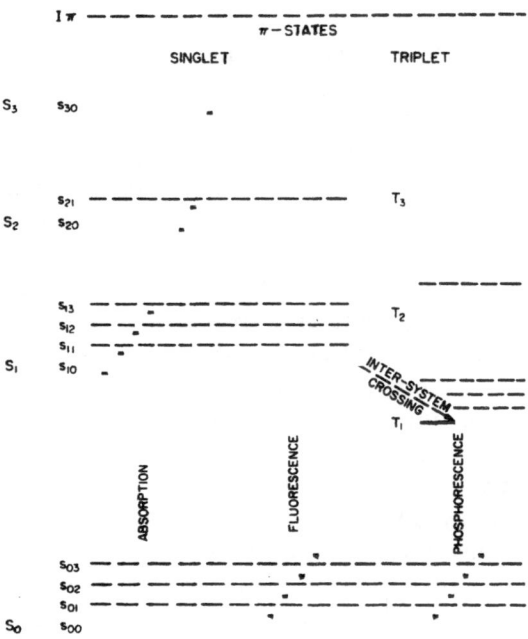

Figure 4-5. Energy levels of π electrons in organic scintillant. [With permission from: J. B. Birks, *Theory and Practice of Scintillation Counting*, p. 47, Pergamon Press, 1967.]

the six π electrons in benzene) are completely delocalized, i.e., they are not associated with any particular atom. Energy levels for both singlet and triplet states of these electrons are shown in Fig. 4.5. The dashed lines represent the vibrational levels superimposed on each electronic level. Absorption of energy takes the electron from the ground state to any one of the excited singlet levels. Transitions from states above S_1 to the S_1 state take place through rapid ($\sim 10^{-11}$ sec) radiationless internal conversion. Transition from the S_1 state to the ground state (to any one of the vibration substates) is accompanied by the emission of fluorescence radiation. The lifetime of the S_1 state is $\sim 10^{-8}$–10^{-9} sec. This is long compared to the period of molecular vibrations ($\sim 10^{-12}$ sec) and therefore the molecule has sufficient time to reach thermal equilibrium before emission. Hence the transition is from the lowest vibration state. Therefore the fluorescence spectrum, decay time, and quantum efficiency are always the same irrespective of the excited level. Also, from Fig. 4.5 it is clear that the absorption spectrum and the fluorescence spectrum are sufficiently separated in wavelength to make these materials good scintillants. A fraction of the molecules in the S_1 state undergo transition to the T_1 state: T_1 states are of long lifetime and transitions from the T_1 state to the T_0 state are responsible for phosphorescence.

The incident ionizing radiation expends part of its energy in raising the π electrons to excited levels. In contrast to inorganic scintillants, where the energy is first given to electrons of alkali iodides and then transferred to activator atoms, the process in organic scintillants is a direct one. As in the case of alkali iodide scintillants, total light output and decay constant vary with temperature.

In liquid solution and plastic solution scintillants, the incident radiation expends its energy in exciting the solvent atoms of the solution. Then the excited solvent atoms transfer their energy to the solute molecules. In turn, the solute molecules emit photons during their transition to the ground state. Thus the emission spectrum is the spectrum of the solute molecules. In this type of scintillant, a good solvent is one that effectively transfers energy to the solute. As in the case of the preceding scintillants, light output and decay constant depend on temperature and concentration.

A few of the major differences between organic and inorganic scintillants are: (1) Pulse height (light output) is much higher in inorganic than in organic crystals. (2) Organic scintillants have a nonlinear relationship between pulse height and energy for charged particles. (3) The light output per MeV for various charged particles differs significantly for organic scintillants. In general, the pulse height per MeV for alpha particles is much less than that for electrons. This ratio of pulse heights is referred to as the alpha-to-beta ratio, which is about 0.1 for anthracene, stilbene, and a number of plastic scintillants.

Table 4.1. Inorganic Scintillants

Material	Wavelength of maximum emission, Å	Decay constant, μsec	Density, g/cm³	Relative pulse height[a]	α/β	Comments
NaI(Tl)	4100	0.25	3.67	210	0.5	Large clear crystals, hygroscopic; must be sealed in air-tight can; used for gamma and charged particle detection and spectroscopy
CsI(Tl)	4200–5700	1.1	4.51	55	~0.8	Available as large, clear, nonhygroscopic crystals; higher detection efficiency for gamma and charged particle detection and spectroscopy.
KI(Tl)	4100	1.0	3.13	~50	1.0	Large, clear, nonhygroscopic crystals; good for detection and spectroscopy of charged particles

LiI(Eu)	4400	1.4	4.06	74	$n/\beta = 0.93$	Large, clear, hygroscopic crystals, available as enriched LiI(Tl); subject to radiation damage from excessive exposure to slow neutrons and ultraviolet light; high-efficiency detector for slow neutron detection and spectroscopy of fast neutrons
ZnS(Ag)	4500	10.0	4.09	~200		Available only as small crystals, transparency to its emitted light is not very good; used for detection of alpha and fission fragments; plastic suspensions are used for neutron detection; not generally used for electron detection; scintillation efficiency for alpha is found to be twice as high as that for electrons

[a] With 10 μsec anode time constant.

Table 4.2. Organic Scintillants

Material	Wavelength of maximum emission, Å	Decay constant, μsec	Density, g/cm³	Relative pulse height[a]	α/β	Comments
Anthracene	4400	0.032	1.25	100	0.1	Can be obtained in a variety of sizes and shapes; most commonly used organic scintillant
trans-Stilbene	4100	0.006	1.16	60	0.1	Available as clear, colorless single crystals up to several inches; crystals are sensitive to thermal and mechanical shocks
Plastic phosphors[b]	3500–4500	0.003–0.005	1.06	28–48	~0.09	Easy to shape, can be obtained in any shape and size
Liquid phosphors[c]	3500–4500	0.002–0.008	0.86	27–49	0.09	Good for 4π counting; most of them can be used in Lucite containers

[a] With 10 μsec anode time constant.
[b] Typical plastic scintillants: polystyrene (solvent) with TP (36 mg/liter) or TPB (16 mg/liter). Polyvinyl toluene (solvent) with TP (30 mg/liter) and αNPO (1 mg/liter) or POPOP (1 mg/liter). A second solute is added to shift the wavelength of light emitted to match the photomultipliers. TP, p-terphenyl; TPB, 1,1',4,4'-tetraphenyl butadiene; POPOP, 1,4-di-[2-(5-phenyloxazolyl)]-benzene.
[c] Typical liquid scintillants: Toluene (solvent) with diphenyl anthracene (10 mg/liter), POPOP (0.5 mg/liter), p-xylene (solvent) with PBD.

Properties of the most commonly used scintillants are given in Tables 4.1 and 4.2.

4.3. *PHOTOMULTIPLIERS*

A photomultiplier is one of the most essential parts of a scintillation detection system. Its main functions are to convert the light signal from the scintillant to an electrical signal and to act as a linear amplifer of $>10^6$ amplification. It should not introduce appreciable time or energy spread.

Typical photomultipliers are shown in Fig. 4.6. The first and most important element is the photocathode, which converts part

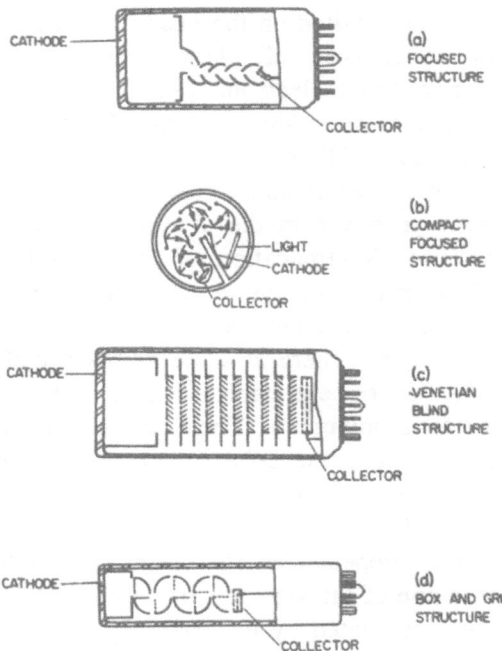

Figure 4-6. Typical photomultiplier designs. [With permission from: T. Sharpe, Electronics Technology, EMI Electronics Ltd., Rpt CP5306 (1961).]

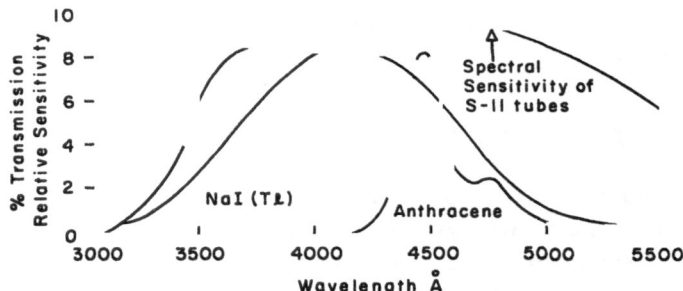

Figure 4-7. Optical sensitivity of S-11 photomultipliers and spectral distribution from typical scintillants.

of the incident photon energy into electrons. In most of the modern tubes designed for scintillation counting, a semitransparent layer of CsSb is deposited on the inside surface of the glass or quartz window. This type of photocathode has about 10% quantum efficiency (photoelectrons per incident photon). The peak sensitivity occurs near 4400 Å. From Fig. 4.7, one can see that these (so-called S-11) photocathodes have good spectral match with most of the scintillants in use.

The electrons produced in the photocathode are accelerated toward the dynodes. Multiplication of electrons takes place in each dynode. The multiplication m_i at each dynode is roughly proportional to the voltage between the dynodes and the total multiplication is $\Pi_{i=1}^{i=k}m_i$, where k is the number of dynodes. The power supply has to be extremely well regulated to keep the variation in m_i to a minimum. The total amplification is also greatly affected by the presence of even small magnetic fields, such as that of the earth. This effect can be reduced by using mu-metal magnetic shields around the tubes. For most photomultipliers the total gain depends on the count rate. The RCA 5819 photomultiplier is one of the very few exceptions.

The dynodes are coated by materials that produce secondary electrons when high-energy electrons strike them. The phenomenon of secondary electron emission is somewhat similar to the photoelectric effect, the difference being that in secondary emission the escaping electrons are produced by an energetic electron rather than by a photon. Therefore photoelectric material can be used for dynode coating. Some of the materials used for coating dynodes are CsSb, AgMgOCs, and MgO. The shape of the dynodes affects the total transit time and the spread in transit time. The spread in transit time is the width at half-maximum of a pulse produced by a single electron. Spread in transit time is the shortest for focused type and compact focused type structures. A large area of the dynodes in unfocused structures enables efficient and uniform electron collection by the first dynode. Focused types are preferred for fast timing experiments, while unfocused types are preferred in other scintillation experiments.

Characteristics of some of the commonly used photomultipliers are given in Table 4.3.

In addition to the photoelectrons emitted by the photocathode, there is also thermionic emission of electrons. The work function of the CsSb photocathode is about 1.9 eV, with a wavelength threshold of 6600 Å. At room temperature the number of electrons having thermal energy above 1.9 eV, that is, the number that can escape the material, is between 10^2 and 10^4 per cm^2 per sec, depending on the way the surface has been prepared. These electrons produced by thermionic emission are also multiplied and are responsible for the dark current, that is, the current produced in the photomultiplier tube in the absence of incident radiation.

In operating the photomultiplier tube for detecting low-energy radiation, it is necessary to reduce the dark current. One technique for the reduction of the dark current is to operate the tube at low temperatures, although even at low temperatures the number of photoelectrons resulting from the thermionic emission is as great as or greater than the number of photoelectrons produced by beta rays of low energy from ^3H or ^{14}C sources.

For purposes of comparison, let us consider a liquid scintillant beta counting system for ^3H. In this case, the maximum energy of the emitted electrons from ^3H is 18 keV with a most probable value around 10 keV. The number of photoelectrons resulting from the absorption of 10-keV beta particles in the liquid scintillant (just before multiplication) is given by the number of

Table 4.3. Characteristics of Representative Photomultipliers[a]

Type	Manufacturer	Photocathode size (min. diam.), in.	Spectral class	Average sensitivity, μA/lumen	Gain	Diameter, in.	Overall dimensions (length), in.
6342-A	RCA	1.5	S-11	50	2.3×10^6 [b]	2¼	5 13/16
6655-A	RCA	1.7	S-11	50	2.3×10^6 [b]	2¼	5 13/16
6810-A	RCA	1.7	S-11	60	6.3×10^8 [c]	2⅜	7½
6199	RCA	1.24	S-11	45	2.8×10^6 [b]	1 9/16	4 9/16
6363	Dumont	2.5	S-11	60	2×10^6 [d]	3	6⅛
6364	Dumont	4.2	S-11	60	2×10^6 [d]	5¼	7⅛
6291	Dumont	1.25	S-11	60	2×10^6 [d]	1½	4¼
9536B	EMI	1.75	S-11	50	6×10^5 [b]	2 1/16	5¾
9578B	EMI	2.5	S-11	50	2×10^6 [b]	3 3/32	6⅛

[a] A more extensive list of photomultipliers and their properties can be found in Ref. 4.
[b] 1250 V supply voltage.
[c] 2300 V supply voltage.
[d] At 145 V/stage.

photons produced in the scintillant × light collection efficiency × quantum efficiency. For a 10-keV beta particle, the number of photoelectrons is

$$(10^4 \text{ eV}/10^2 \text{ eV}) \times 10\% \times 10\% = 1$$

It is evident that the production of photoelectrons from the beta particle is much less than that from thermionic emission. Therefore it appears to be difficult to distinguish between the signal produced by a thermionic electron and the signal produced by the incident radiation.

Figure 4.8 shows a coincidence counting system which can be used to reduce the effect of thermionic emission. A photomultiplier tube is placed on either side of the liquid scintillant. The signals from the photomultipliers are fed into a coincidence system. Since the thermionic emission is random it will be highly improbable to get coincidence due to thermionic electrons from both tubes. Therefore the coincidence output will be due to the

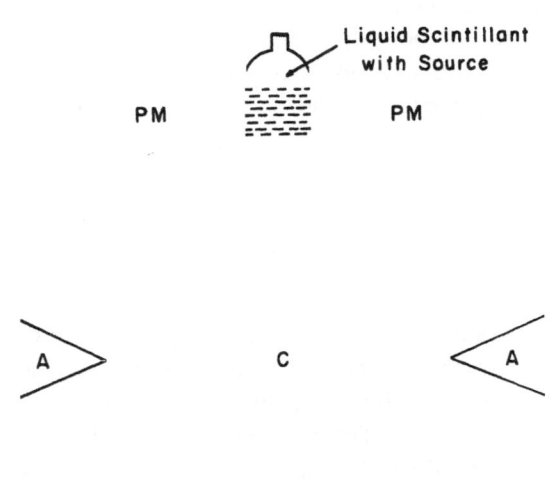

Figure 4-8. Coincidence system for measurements using liquid scintillants.

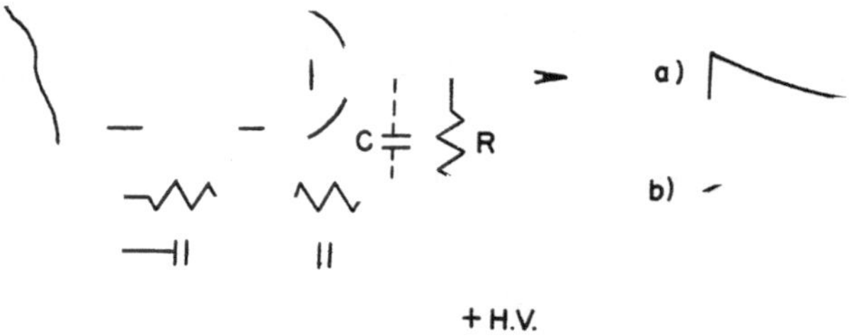

Figure 4-9. Photomultiplier anode circuit. Waveforms (a) and (b) depict output pulses from fast $(\tau \ll RC)$ and show $(\tau \gg RC)$ scintillants, respectively.

incident radiation. The photomultipliers for such a system are operated at low temperature.

Liquid scintillants are used for counting low-energy beta rays because with such a scintillant, the radioactive sample can be introduced into the liquid medium and thereby the losses that result in solid scintillants can be eliminated. For solid scintillants, the source has to be kept outside the case surrounding the scintillant and, as a consequence, there are radiation losses by absorption in the container and backscattering from the case and the scintillant.

4.4. THE SHAPE OF THE PULSE

There are two factors controlling the shape of the pulse from the scintillation counter: (1) the decay time of the scintillant and (2) the time constant of the collector resistance R and the stray capacitance C (Fig. 4.9). The number of electrons across the stray capacitor will depend on the rate at which they reach the collector and the rate at which they discharge through the resistance. The problem, therefore, is similar to the series decay in radioactive materials discussed in Chapter 2. If n_p is the total number of atoms in the scintillant, the number of electrons reaching the collector is

$$n_p\epsilon_c\epsilon \prod_{i=1}^{k} m_i e^{-t/T_d}$$

where ϵ is the number of electrons produced at the photocathode per photon falling on the photocathode and ϵ_c is the collection efficiency of the photocathode, which includes the effects of absorption of photons by the scintillant. The quantity $\prod m_i$ is the product of the multiplications at all the dynodes. We are assuming that there is no time spread in the photomultiplier. The rate of decay is determined by the term $e^{-t/RC}$. Therefore the equation for the number of electrons on the capacitor is

$$n(t) = n_p\epsilon_c\epsilon \prod_i m_i \left(\frac{RC}{RC - \tau}\right)(e^{-t/RC} - e^{-t/T_d})$$

Dividing this equation by C and multiplying by e, one gets the equation for the voltage, namely

$$V(t) = \frac{n_p\epsilon_c\epsilon \prod_i m_i e}{C}\left(\frac{RC}{RC - \tau}\right)(e^{-t/RC} - e^{-t/T_d}) \qquad (4.3)$$

The amplitude of the pulse depends on the decay constant of the scintillant and the time constant RC. If $RC \gg \tau$, the pulse amplitude is independent of τ [Eq. (4.3) for $RC \gg \tau$]. This means that the emission process is completed in a time much shorter than the integration time of the circuit and the amplitude is a maximum when this condition is satisfied. Figure 4.9 shows two typical pulses for a fast (small τ) and a slow (large τ) scintillant.

4.5. SCINTILLATION COUNTER ASSEMBLY

A scintillation counter assembly for a solid scintillator is shown in Fig. 4.10. The crystals can be bought as a separate unit in which they are sealed in an aluminum can containing reflectors and an optical window. A problem may arise in attaching the crystal to the photomultiplier. To get good optical coupling, Dow Corning

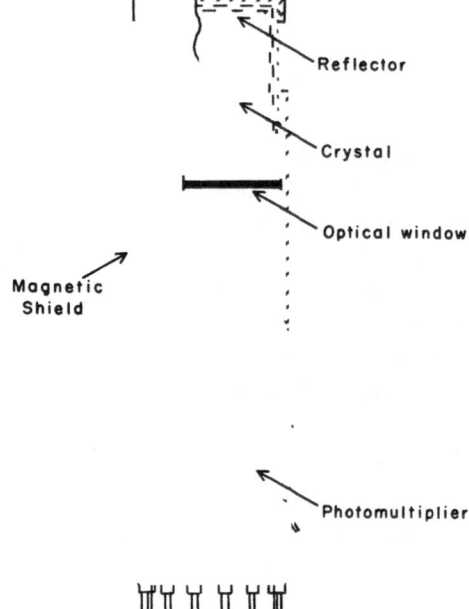

Figure 4-10. Scintillation counter assembly.

Silicone 200 fluid, Dow Corning e-2-0057 silicone compound, or ophthamological petroleum jelly can be used. The last two compounds are gelatinous and therefore are easier to use.

The voltage distribution to the different electrodes is achieved through a voltage dividing system, as shown in Fig. 4.11.

4.6. *RELATIONSHIP BETWEEN PULSE HEIGHT AND ENERGY*

For spectroscopic studies it is advisable to have some known relationship between the energy of the radiation and the height of the pulse produced by that radiation. For gas counters, except those operating in the Geiger region, and for semiconductor counters (discussed in the following chapter) this relationship is linear. In scintillation counters, the behavior of different scintillants varies in this respect.

For inorganic scintillants over a wide range of energy, the pulse height is linearly proportional to the energy for electrons, protons, and deuterons incident upon the scintillant (nonlinearity produced by backscattering for β rays is discussed below). In the case of gamma rays, scintillants receive energy from the electrons produced by gamma rays, and therefore the same linear relationship holds between gamma-ray energy and the pulse height. Slight nonlinearity has been observed for electrons produced by gamma rays below a few hundred keV. For heavier particles, nonlinearity occurs over a wide range of energy. In the case of organic scintillants, nonlinearity for beta and gamma rays occurs at very low energies, approximately a few keV, while nonlinearity for heavier particles extends over a wider range of energy.

Another point to consider is the relationship between the pulse height and the nature of the particle. The height of the pulses produced by heavily ionizing particles such as alpha

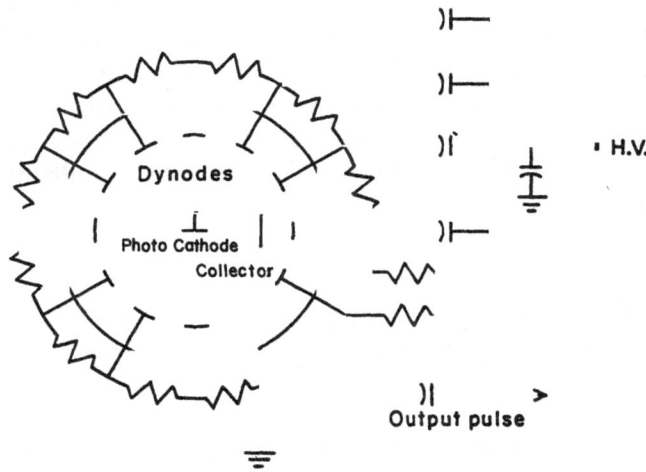

Figure 4-11. Voltage distribution to electrodes of the photomultiplier.

particles may differ considerably from that produced by electrons of the same energy. This difference depends on the nature of the counters, being generally small in gas counters and semiconductor counters. The energy required to produce an electron–hole pair in Si for alpha and beta particles is 3.62 and 3.67 eV, respectively, so that the difference here is of the order of 1%. Similar results have been obtained for Ge detectors. The energy per ion pair for alpha and beta particles is slightly different for all the commonly used counter gases. Beta particles have to expend 42.3 eV and alpha particles 42.7 eV to produce an ion pair in helium; the corresponding values are 27.3 and 29.2 eV in CH_4. However, this difference is much higher in scintillation counters, and it is usually given as the α/β ratio, that is, the ratio of pulse height produced by an alpha particle to that produced by a beta particle of the same energy. For NaI(Tl), α/β is 0.5; for anthracene, 0.1; and for a plastic scintillant, 0.09 (see Tables 4.1 and 4.2).

In studies involving beta rays, the pulse height distribution is distorted due to backscattering, especially for energies <150 keV. Radiation is sometimes scattered out of the counter before all its energy is deposited. This tends to increase the number of pulses recorded in the low-energy region. The scattering increases exponentially with the atomic number Z of the scatterer and also increases with decreasing energy. Therefore counters of low-Z materials, such as organic compounds for scintillants and silicon for semiconductor counters, are preferred. While NaI scatters almost 90% of the incident beta particles before they are completely stopped, anthracene scatters only 8%. Special arrangements, such as locating the source in a split scintillant crystal, reduce the effect.

4.7. DETECTION OF GAMMA RAYS BY SCINTILLATION COUNTERS

Gamma rays are stopped in the scintillant through one of the following processes (see Chapter 2): (1) the photoelectric effect,

(2) the Compton effect, and (3) pair production. Electrons are produced in the first two processes and electrons and positrons in the third. These charged particles excite the scintillant, producing photons, and therefore the pulse height produced by the gamma ray is proportional to the energy of the electron (and positron). Hence the pulse height distribution (that is, the number of pulses versus pulse height) depends on the relative cross sections for these processes.

In order to explain the pulse height distribution, let us consider a radioactive source emitting 2-MeV gamma rays and limit ourselves to the first two processes. The photoelectrons have energy of 2 MeV, whereas the Compton electrons exhibit a continuous energy distribution ranging from 1.775 MeV at $\theta = 180°$ to 0 MeV at $\theta = 0°$, where θ is the Compton angle. Consequently, the pulse height distribution in an ideal case (infinite resolution) is as shown in Fig. 4.12. The ordinate of point 1, representing number of pulses produced by photoelectrons, and the area under the curve 2, representing the number of pulses produced by Compton electrons, are proportional to their respective cross sections for the 2-MeV gamma ray. The vertical line at

ı

2

a)

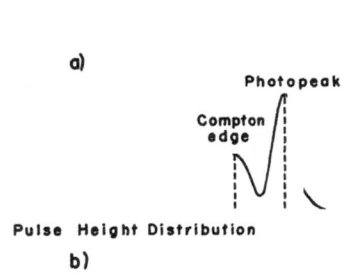

b)

Figure 4-12. Pulse height distribution for a single-energy gamma source. (a) Infinite resolution. (b) Finite resolution.

Table 4.4. Contributions to Pulse Height Distribution due to Events Occurring in the Scintillant[a]

	A	B	C	D	E	F	G
Absorption process	Photoelectric effect	Compton effect (scattered gamma escapes)	Compton effect; scattered gamma is absorbed	Pair production annihilation, gamma rays escape	Pair production, one of the annihilation radiations escapes	Pair production, both radiations absorbed	Pair production, one of the annihilation radiations produces (Compton) electron
Energy absorbed in the crystal	E_γ	Varies from 0 to T_{max}; $T_{max} = E_\gamma \times \left[1 + \dfrac{m_0c^2}{2E_\gamma}\right]^{-1}$	E_γ	Kinetic energy of the electron–positron pair $= E_\gamma - 1.02$ MeV	$E_\gamma - 1.02 + 0.51 = E_\gamma - 0.51$	E_γ	Compton contribution + $E_\gamma - 0.51$
Pulse height	Pulses in the photopeak	Pulses in the Compton distribution	Pulses in the photo peak	Proportional to $E_\gamma - 1.02$ MeV and the peak is called double escape peak	Proportional to $E_\gamma - 0.51$ MeV and peak is called single escape peak	Photopeak	Pulses in the Compton continuum

[a] E_γ is the energy of the incident gamma ray. In the photoelectric effect, the electron produced takes all the energy of the gamma ray. In the Compton effect, the energy gained by the electron varies from 0 to T_{max}, depending on the angle of scattering. The electron gains T_{max} when the gamma is scattered back (scattering angle π radians). T_{max} is given by the equation in column B, where m_0 is the rest mass of the electron, and m_0c^2 is equal to 0.51 MeV. In pair production, 1.02 MeV is converted to the mass of the electron–positron pair. In the opposite process, where the positron annihilates with an electron, two gamma rays of 0.51 MeV each are produced. In some cases—proportional counters, thin scintillants, and Si(Li) detectors—x rays produced during the photoelectric effect escape. This will produce an additional peak, on the low-energy side, with pulse height proportional to $E_\gamma - E_x$.

Table 4.5. Contributions to the Gamma Spectrum by Gamma Rays Scattered from the Source and Surroundings into the Scintillant

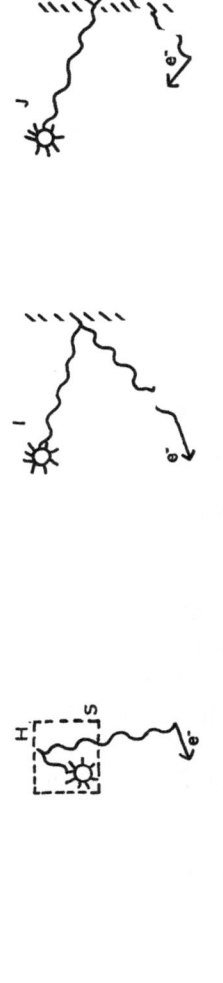

Event	Compton backscattering	Compton at angle other than π	Pair production in the environment and one gamma scattered into the scintillant
Energy	$E_\gamma - T_{max}$ is the maximum energy of the electron[a]	Energy between $E_\gamma - T_{max}$ and E_γ	0.51 MeV
Pulse height	Peak due to backscattered gamma, called backscattered peak, occurs at a well-defined position with respect to the photopeak	Pulse height distribution indistinguishable from the regular Compton distribution	Peak at 0.51 MeV; also a peak at this position if the source decays by positron decay

[a] See footnote to Table 4.4.

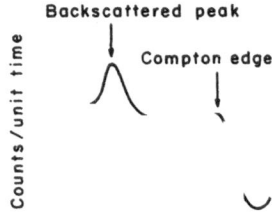

Photo peak

Backscattered peak

Compton edge

Counts / unit time

Pulse height

Figure 4-13. Pulse height distribution for gamma rays from ^{137}Cs.

the end of curve 2 is called the Compton edge, which corresponds to the maximum energy of the Compton electrons.

The finite resolution of the scintillant system introduces a spread in the pulse height distribution. As a result, a single, continuous curve with two peaks is obtained. The first peak is called the photopeak and the second peak is the Compton edge.

In Table 4.4, all contributions to the pulse height distribution due to events occurring in the scintillant are discussed. Notice that the effects of pair production are also included here. Other factors, such as scattering into the scintillant, could also occur. Such possibilities are indicated in Table 4.5.

As examples of pulse height distributions, Figs. 4.13 and 4.14 illustrate the distributions for ^{137}Cs and ^{24}Na. For ^{137}Cs we have contributions from A, B, and C, and possibly from H and I, of Tables 4.4 and 4.5. In the distribution for ^{24}Na, contributions from the pair production events (D and E of Tables 4.4 and 4.5) can be identified. Pulses due to events C, F, and G fall in photopeak A and those due to I fall between E_γ and $E_\gamma - T_{max}$.

The ratio of number of counts under the photopeak to the

total number of counts increases with increase in the size of the counter. If multiple processes are not taking place, this ratio will be equal to the ratio of the photoelectric cross section to the sum of the photoelectric, Compton, and pair production cross sections. However, as the size of the crystal increases, the chance becomes greater for multiple processes, increasing the peak-to-total ratio.

4.8. INTEGRAL COUNTING

From Figs. 4.13 and 4.14 it is clear that the number of pulses produced by a gamma ray of a given energy varies from 0 to a maximum. Therefore, if one is interested only in determining the number of gamma rays falling on the crystal, all the pulses should be counted. However, the electronics only allow us to count pulses above a certain height, and sometimes the noise in the electronics determines this lower limit. Counting of pulses above a certain voltage is sometimes known as integral counting, as

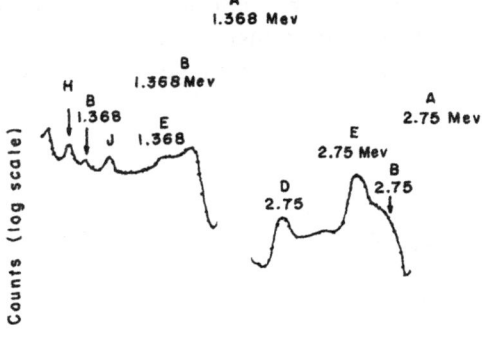

Figure 4-14.   Pulse height distribution for gamma rays from ^{24}Na.

H.V.

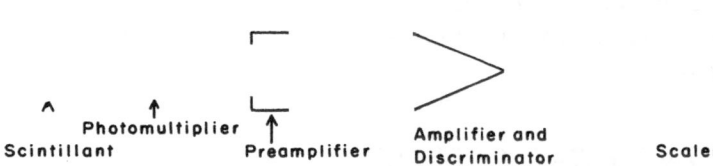

Scintillant   Photomultiplier   Preamplifier   Amplifier and Discriminator   Scaler

Figure 4-15. Block diagram of an integral counting system.

opposed to differential counting, where pulses between two fixed heights (a lower limit and an upper limit) are counted.

A block diagram of the electronics used in this type of counting is shown in Fig. 4.15. The minimum height of the pulses counted is determined by the sensitivity of the scaler, and the height of the pulses reaching the scaler depends on the voltage applied to the photomultiplier and the linear amplifier. Sometimes the amplification is fixed in the system. In such systems part of the gamma spectrum counted and hence the efficiency of the system depend only on the voltage. The shape of a characteristic curve of counts versus applied voltage will also depend on the

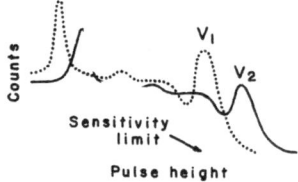

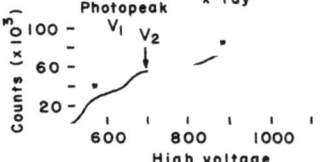

Figure 4-16. The effect of applied voltage on the change in total counts. As the voltage increases, the pulse height distribution moves to the right and therefore number of pulses with height above the "sensitivity limit" of the scaler increases.

radioactive source. Let us consider a single gamma source such as ^{137}Cs. If none of the pulses have height equal to the sensitivity (see Fig. 4.16), we will not get any counts. As the voltage is increased, the spectrum stretches toward the higher pulse height side and more and more pulses will be counted. For cesium a small "plateau" is seen for voltages between V_1 and V_2 where the count increase is very small. Also, changes in slope occur when peaks cross the sensitivity limit. From the above discussion, it can be seen that the counter efficiency depends on the voltage.

4.9. *DIFFERENTIAL COUNTING AND DETERMINATION OF ENERGY OF GAMMA RAYS*

In differential counting one is interested in finding the number of pulses for different pulse heights. The instruments needed for

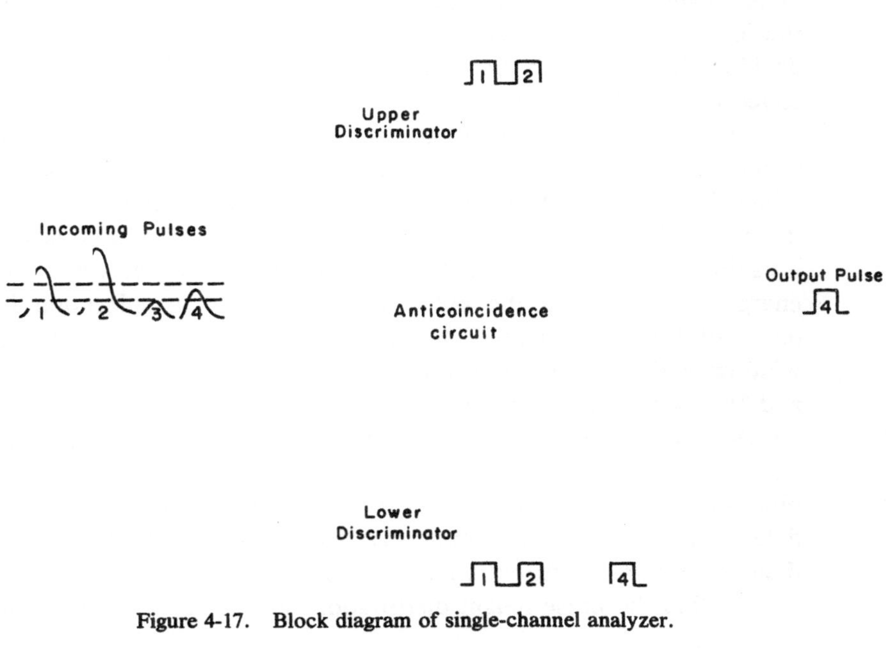

Figure 4-17. Block diagram of single-channel analyzer.

such an analysis are shown in Fig. 4.1. The pulses from the photomultiplier are amplified and fed into a single-channel or a multichannel analyzer.

The single-channel analyzer can be explained with the aid of Fig. 4.17. The pulses go into an upper level and a lower level discriminator. The discriminators give output pulses only for those pulses having height higher than the setting on the discriminator. The upper level discriminator gives output pulses for input pulses 1 and 2. The lower level discriminator gives output pulses for input pulses 1, 2, and 4. These pulses go into an anticoincidence circuit. This circuit gives an output pulse only for those pulses from the lower discriminator unaccompanied by pulses from the upper discriminator, that is, for input pulse 4. If pulses from both discriminators come into anticoincidence at the same time, they will be cancelled. The upper and lower discriminators are set by externally controlled potentiometers. The potentiometers are usually connected internally so that the upper voltage is equal to the lower voltage level plus the voltage on the upper level potentiometer. Therefore the upper level changes as the lower level is adjusted. The setting on the upper level, which decides the difference between the upper and lower levels, is called the window width.

In an actual experiment the pulse height, even though it is proportional to the energy absorbed in the scintillant, is determined by the voltage on the photomultiplier and amplification in the preamplifier and the linear amplifier. Therefore for each setup a calibration is needed to determine the relationship between the energy absorbed and the pulse height. Usually a calibration is obtained by taking counts for different lower levels for fixed window width and plotting pulse height versus counts for ^{137}Cs and ^{60}Co sources. From these plots the pulse height is determined for the photopeaks corresponding to the energies 0.66, 1.1, and 1.3 MeV, and a pulse height versus energy plot is obtained. This plot will be a straight line. This calibration can be used to determine the energy of gamma rays from unknown sources once their pulse height distribution is obtained.

To find the pulse height distribution, pulses can be fed into a multichannel analyzer. The operation of the multichannel analyzer can be understood if we assume that we have a series of discriminators and anticoincidence circuits as shown in Fig. 4.18. A pulse is obtained from the ith anticoincidence circuit for an input pulse of height between V_i and V_{i+1} volts and it is stored in the ith magnetic core (that is, a memory core). This system therefore has the advantage that the pulses are analyzed as they come in and the number of the pulses is stored in the proper channel according to their height. In each single channel the pulses between the upper and lower discriminator settings are counted at one time. After a predetermined time of analysis, one obtains from the memory the numbers of the stored pulses in each channel. Then one can plot channel number versus counts and from this plot one obtains the channel number at which the maximum of photopeaks of known energy occurs. From this a

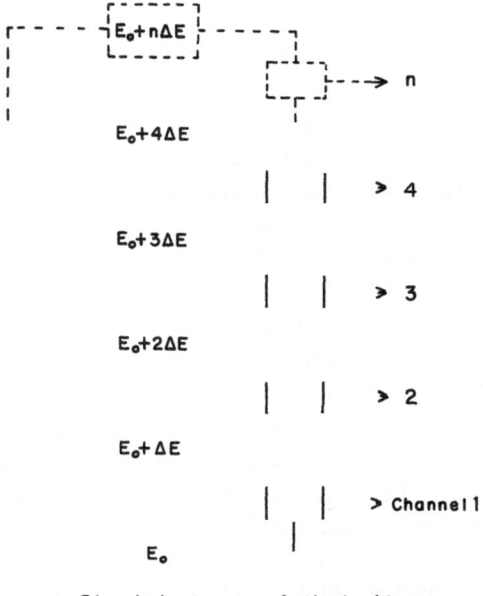

Figure 4-18.   Multidiscriminator pulse height analyzer system.

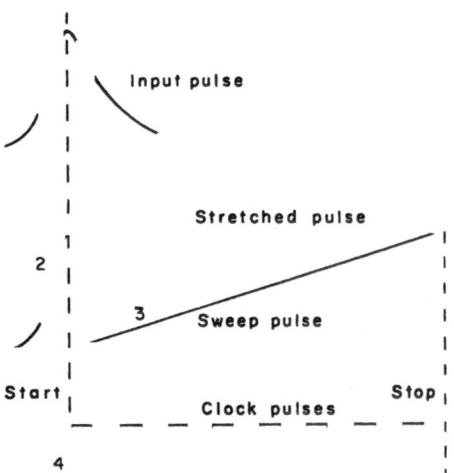

Figure 4-19.   Principle of pulse height-to-time
conversion.

plot of energy versus channel number can be made to calibrate
the system.

The system discussed above is called a multidiscriminator.
The multidiscriminator pulse height systems are not widely used,
because of two important disadvantages. First, the number of
components is too high since each channel has a discriminator
and an anticoincidence circuit. This increases the initial cost.
Second, it is difficult to adjust the channels to be equal in width,
and because of inevitable drifts, it is difficult to keep them in
adjustment. However, the time required to analyze a pulse is very
short for these systems compared to modern analyzers.

Modern pulse height analyzers work on the basis of pulse
height-to-time conversion first suggested by Wilkinson. Figure
4.19 illustrates the principle. The pulse-height-to-time converter
stretches the input pulse (1). The height of the stretched pulse (2)
is equal to the maximum height of the input pulse and it is
maintained constant. At the time t_1, when the input pulse attains
its maximum value, two other pulses are initiated. The first pulse
(3) is a linearly increasing sweep pulse and the second (4) is a

series of timing pulses. When the sweep pulse reaches the height of the stretched pulse at time t_2, the timing pulses stop. The number of timing pulses in the time interval $t_2 - t_1$ is proportional to the pulse height. Information is usually stored in the memory, given by the channel number, as decided by the pulse height or the number of timing pulses. Information can be obtained from the memory in the form of channel number and number of pulses in each channel.

4.10. EFFICIENCY OF SCINTILLATION COUNTERS FOR GAMMA DETECTION

Scintillation counters are about 100 times more efficient than gas counters for detection of gamma rays. The reason for this is that the gamma ray will encounter more "material" in a scintillation counter. The number of gamma rays stopped will depend on the energy of the gamma ray since the absorption cross section varies with the energy. If the gamma rays are incident perpendicular to the face of the crystal, the number of gamma rays stopped in the crystal, or detected by the scintillation counter, is given by the equation

$$N_0 - N = N_0(1 - e^{-\mu x}) \qquad (4.4)$$

where μ is the absorption coefficient of the gamma ray, x is the thickness of the crystal, N_0 is the number of gamma rays incident on the crystal, and N is the number passing through the crystal. From Eq. (4.4), the efficiency of the counter is given by

$$[(N_0 - N)/N_0]100 = (1 - e^{-\mu x}) \times 100\% \qquad (4.5)$$

If the gamma source is a point source at a distance h from the face of the crystal, as shown in Fig. 4.20, the equation for efficiency becomes

$$\epsilon = \frac{\int_0^{\Omega_0} (1 - e^{-\mu x})\, d\Omega}{\Omega_0}$$

where Ω_0 is the solid angle subtended by the front face of the crystal at the source. Here we are calculating an average

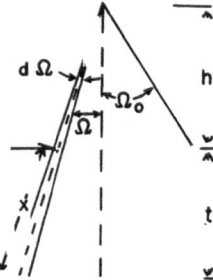

Figure 4-20. Location of a gamma source from a
crystal face.

efficiency since the distance x traveled by the gamma ray is a
function of Ω. Also, x is a function of the distance h of the source
from the crystal and the thickness t of the crystal. The variation
of efficiency with energy for different values of h is shown in Fig.
4.21 for a crystal of 2 in. diameter and 1 in. length. For other
crystal sizes, see Ref. 3.

The efficiency of the counter will be less than that given by
Eq. (4.5) since the discriminator of the electronic scaler rejects
some of the smaller pulses. Hence the efficiency depends on the
discriminator level of the scaler, which is very difficult to
determine. This difficulty can be avoided by using a single-
channel analyzer to count the pulses in the photopeak. The ratio
of counts in the photopeak to the number of gamma rays of
energy equal to the photopeak energy incident on the crystal is
defined as the intrinsic photopeak efficiency. Intrinsic peak
efficiency is the probability that a gamma ray of given energy will
produce a pulse falling in the full energy peak. Figure 4.22 shows
the intrinsic peak efficiency for various crystal sizes as well as for
different source-to-crystal distances. Intrinsic peak efficiency
increases with the size of the crystal. This is because the chance
of the crystal receiving the total energy by multiple processes
increases with the size of the crystal.

For gamma-ray detection, NaI(Tl) crystals are usually used
because of their higher density and resultant higher efficiency.
Organic or plastic scintillants can be used for gamma detection,
especially for a fast-rising pulse. Since the density of organic or

plastic crystals is lower than that of the sodium iodide crystal, the efficiency is improved by using larger crystals.

4.11. *ENERGY RESOLUTION IN SCINTILLANT DETECTORS*

Another factor of importance in the detection of gamma rays is the ability of the system to resolve photopeaks that are close

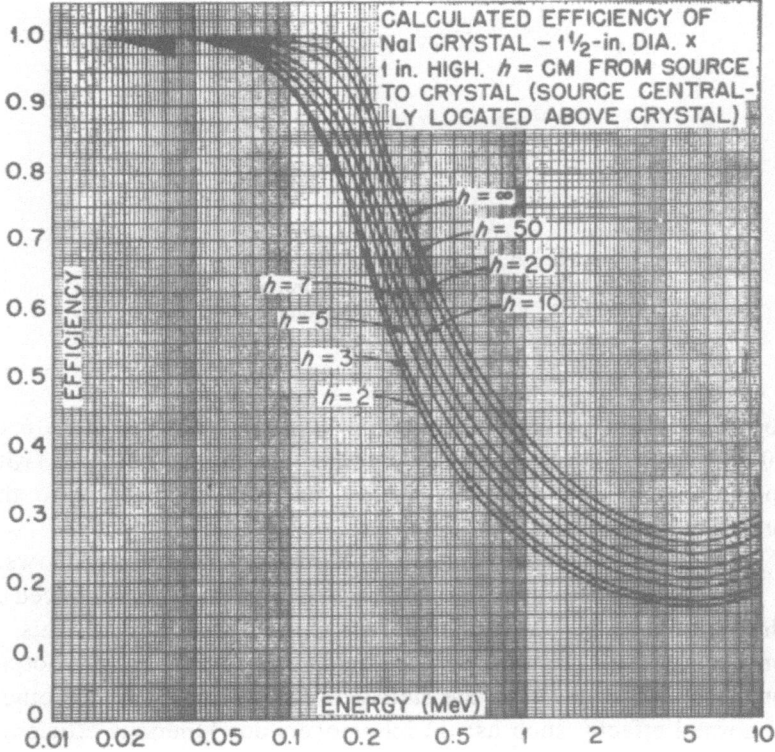

Figure 4-21. Variation of efficiency of gamma rays as a function of energy for different source-to-crystal distances. [With permission from: P. R. Bell, in *Alpha-, Beta-, and Gamma-Ray Spectroscopy* (K. Seigbahn, ed.), North-Holland Publishing Co., Amsterdam, 1965.]

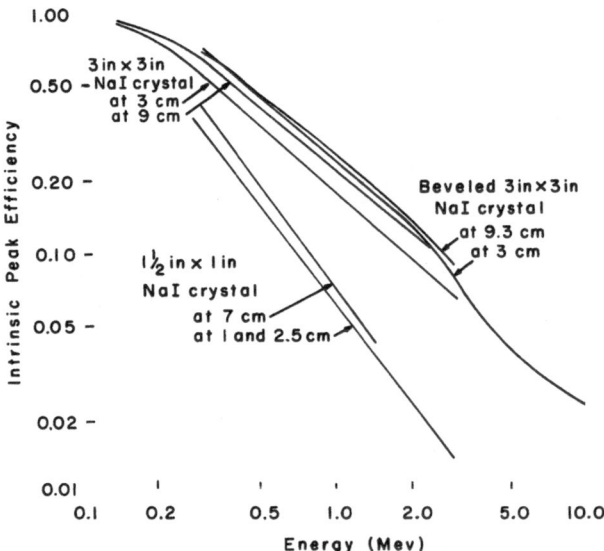

Figure 4-22. Intrinsic peak efficiency of NaI(Tl) detectors as a function of crystal size and source-to-crystal distance. [With permission from: J. H. Neiler and P. R. Bell, in *Alpha-, Beta-, and Gamma-Ray Spectroscopy* (K. Seigbahn, ed.), Vol. 2, Chapter V, North-Holland Publishing Co., Amsterdam, 1965.]

together in energy. For scintillation counters R varies from 6 to 10%, which is not as good as the resolution obtained with ionization chamber counters but is generally comparable to the proportional counters.

Some of the reasons for the higher value of R are as follows.

(1) Statistical variation in the number of photons produced in the scintillant is one major factor contributing to R. This is proportional to $\sqrt{n}$, where n is the average number of photons produced. In addition to the statistical fluctuation there are other, abnormal effects, such as: (a) inhomogeneous luminous efficiency throughout the crystal, mainly due to variations of Tl concentration; (b) nonproportional scintillation response, that is, luminous efficiency varying with specific ionization; and (c) multiple interaction effects.

(2) Variation in the amount of light collected at the photo-cathode.

(3) Variation in the production of photoelectrons. The average efficiency for production of photoelectrons is usually 14–19%. It takes approximately 350 eV to produce one photoelectron. If we consider the first three steps (production of photons, collection of photons, and production of photoelectrons) as one for calculation of the statistical variation in the number of photoelectrons, we obtain $(350/10^6)^{\frac{1}{2}} \times 10^2 \approx 2\%$.

(4) Variation in the collection of electrons at the first dynode.

(5) Variations in the multiplication at successive dynodes.

A good discussion of the effects of these factors can be found in Chapter 5 of Ref. 4. A very simple equation for resolution, due to Neiler and Good,[5] is

$$\Delta E/E = 2.35N^{-\frac{1}{2}}(1 + \beta^{-1})^{\frac{1}{2}} \qquad (4.6)$$

where N is the number of photoelectrons produced by the radiation, ΔE is the half-width for the energy E, and β is related to the gain per stage of the photomultiplier. The quantity β takes care of the contribution to the FWHM given by the variation in amplification at the dynodes. Since N is proportional to the input energy if β is held constant (constant applied voltage), ΔE will be proportional to $E^{\frac{1}{2}}$. The results of a recent study[6] are shown in

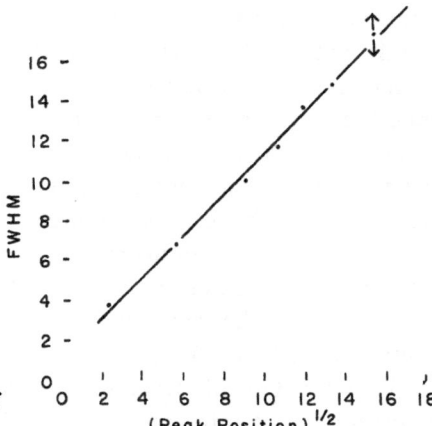

Figure 4-23. Variation of FWHM of photopeak as a function of energy[6]

Fig. 4.23. It may be pointed out here that the conventional plot is resolution in percent vs. (energy)$^{-\frac{1}{2}}$. For the study of Ref. 6, ^{116}In was used. The source ^{116}In can be produced by the ^{115}In(n, γ)^{116}In reaction. This source has several gamma rays, ranging in energy from 0.137 to 2.12 MeV. Therefore, as shown in Ref. 6, the source can be used for (1) calibration of the detector–analyzer system, (2) study of the variation of half-widths of photopeaks in relation to their energy, and (3) variation of the efficiency of the detector as a function of energy. The ^{116}In nucleus ($t_{\frac{1}{2}}$ = 54 min) is often used in undergraduate laboratories as a source in experiments on half-life determinations.

As mentioned earlier, the resolution for energies in the ~1–5 MeV range for NaI is 6–10%. To obtain the best resolution requires careful matching of the photomultiplier sensitivity and the scintillant spectra, proper optical coupling between the crystal and the photomultiplier, and selection of a low-noise photomultiplier, a crystal with uniform Tl distribution, and suitable reflectors.

4.12. DETECTION OF CHARGED PARTICLES WITH SCINTILLATION COUNTERS

Any of the scintillants can be used for detection of charged particles. For alpha particles, since their range is very small, thin crystals are used. Other radiations, such as beta particles and gamma rays, will produce only small pulses in thin crystals, and hence they can be discriminated in alpha detection. As mentioned earlier, it is difficult to grow good crystals of ZnS and they have a very poor light transmission ability. The thin crystals required for alpha detection have good enough transmission, but they do not have a sufficiently uniform light output to make them good detectors in the determination of alpha energies. Also, the long decay time of the scintillants makes them poor detectors for counting large alpha fluxes.

Thin crystals of organic scintillants are hard to handle and too brittle to cut into thin crystals. Light output is also not very high. Thin crystals of activated sodium iodide are therefore the best choice for an alpha-particle spectrometer.

In beta counting with scintillants, special consideration should be given to two factors: (1) backscattering and (2) the continuous energy distribution for beta particles from radioactive sources. Backscattering, as mentioned earlier, is very small for organic and plastic scintillants. Therefore these are used for electron spectroscopy. Various 4π-geometry arrangements, such as sources located between split crystals and sources mixed with liquid scintillants, give very high efficiencies. However, counting low-energy beta particles from ^{14}C (E_{max} = 155 keV) and ^{3}H (E_{max} = 19 keV) is always a problem. As mentioned before, a difficulty in these cases arises from the inability to distinguish between signals due to the few photoelectrons produced by the beta particles and signals due to the thermal electrons from the photomultiplier. Cooling the photomultiplier tubes and looking at the scintillant with two photomultiplier tubes in coincidence are two often used techniques to reduce the noise. There are excellent commercially available liquid scintillant systems utilizing both these techniques which give 90% efficiency for ^{14}C beta particles and 30% efficiency for ^{3}H beta particles.

4.13. *DETECTION OF NEUTRONS*

As discussed in Chapter 3, for the detection of neutrons we depend on a process in which a charged particle is produced. For fast neutrons we depend on n–p elastic scattering and for slow neutrons we depend on a ^{10}B(n, α)^{7}Li or ^{6}Li(n, ^{3}H)^{4}He reaction. Both these reactions have the advantage that no gamma rays are produced and the reaction products have very high kinetic energies, 2.79 MeV for the first reaction and 4.78 MeV for the second reaction.

Lithium iodide crystals activated with Ag or Eu are used in slow neutron detection. A typical pulse height spectrum produced by LiI(Eu) is shown in Fig. 4.24. From this it is clear that with proper discrimination one can count only the thermal neutrons

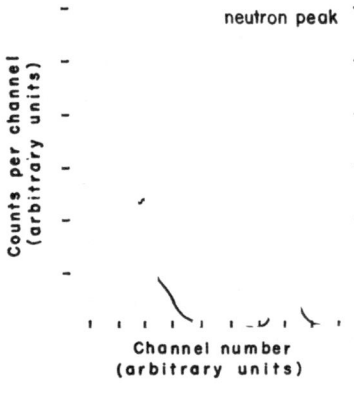

Figure 4-24. Pulse height spectrum produced by slow neutrons using a LiI(Eu) crystal.

since the pulses produced by them form a distinct peak and are well separated from the pulses produced by background radiations or by other reactions in the crystal.

It is of interest to calculate the efficiency of a LiI(Tl) counter. As we have seen, the intensity of neutrons decreases exponentially with thickness.

$$I = I_0 e^{-\mu_l x}$$

where x is the thickness of the absorber and $\mu_l = \Sigma_i N_i \sigma_i$; N_i is the number of different nuclei per unit volume and σ_i is the absorption cross section. The fraction of neutrons absorbed then becomes

$$\epsilon_n = (N_{Li} \sigma_{Li} / \mu_l (1 - e^{-\mu_l x})$$

For europium-activated lithium iodide scintillation counter, the macroscopic cross section μ_l is 1.47 cm^{-1} (for this calculation, normal isotopic composition, 92.5% ^{7}Li and 7.5% ^{6}Li, is assumed) and therefore a crystal thickness of 2–4 cm is large enough to obtain total absorption of thermal neutrons. For a 95% enriched (95% of Li atoms are ^{6}Li isotopes) LiI(Eu) scintillant, $\mu_l = 16$ cm^{-1}, so that only a 0.2–0.5 cm thickness is used for the total absorption of thermal neutrons. A LiI(Eu) scintillant can also be used to detect fast neutrons. Efficiency decreases with energy. For a 10-mm crystal, efficiency is 0.44% for 1-MeV neutrons and is 0.05% for 14-MeV neutrons.

Any organic scintillant can be used as a fast neutron detector. However, because of the large light output for electrons compared to recoil protons, it is impossible to use them in the presence of a large gamma background. A thermal neutron detector can be used as a fast neutron detector if it is covered with a moderating hydrogenous substance such as paraffin or plastic.

4.14. *PARTICLE IDENTIFICATION BY PULSE SHAPE DISCRIMINATION*

The fluorescence emission from some organic scintillants contains a fast component and a slow component. The slow component decays nonexponentially over several microseconds. The fraction of light produced in these two groups depends on the exciting

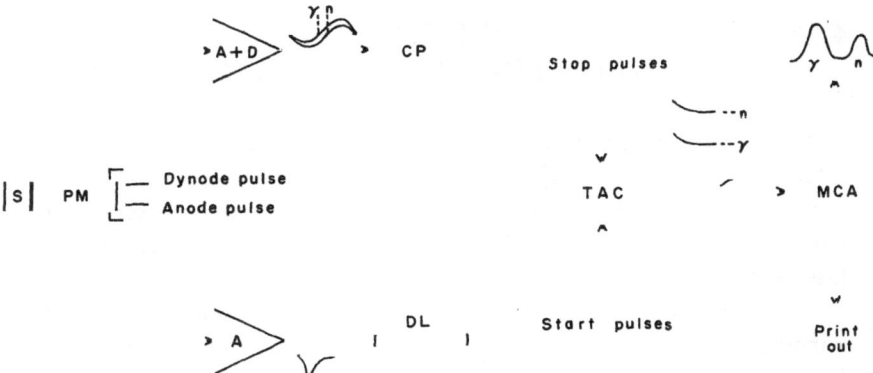

Figure 4-25. Block diagram illustrating the principle of pulse shape discrimination. Neutrons and gamma rays can easily be separated by this method using organic or liquid scintillants. S, scintillant; P, photomultiplier; A + D, amplifier and differentiator; CP, crossover pickoff; TAC, time-to-amplitude converter; MCA, multichannel analyzer.

particle. Therefore the shapes of the pulses produced by different particles vary sufficiently to identify the particles by electronic analysis. Neutron–gamma discrimination based on this principle was suggested and demonstrated by Brooks[7]. *Trans*-stilbene appears to be the most suitable scintillant for pulse shape discrimination; anthracene, *p*-terphenyl, naphthalene, and some of the liquid and plastic scintillants may also be used for this purpose.

Since the shapes of the pulses change with the nature of the incident radiation, when they are amplified and integrated they will have measurable differences in their shapes. These pulses, when doubly differentiated, cross the zero level at distinctly different times from the start of the pulse. These times are measured in terms of the pulse height produced in a time-to-amplitude converter. The start and stop pulses are produced, respectively, at the start of each pulse and at the crossover point (see Fig. 4.25). The shapes of the signals at different points are indicated in Fig. 4.25. This method is especially suitable for distinguishing between neutrons and gamma rays in the same energy range.

4.15. *OTHER METHODS FOR PARTICLE IDENTIFICATION AND DISCRIMINATION*

Detecting certain radiations in the presence of others is often an interesting and challenging problem. We have seen a good example of this in the detection of neutrons in the presence of gamma rays. Other methods for discrimination are discussed in this section.

Discriminating against heavier particles to count only light particles is not very difficult. For example, counting gamma rays in the presence of electrons can be easily achieved by using an absorber of suitable thickness to stop electrons and let only gamma rays into the counter. Similar arrangements can be used to count beta particles in the presence of alpha particles. To achieve the reverse, detecting heavier particles in the presence of lighter ones, requires some special detector designs. These are discussed below.

4.15.1. *Adjusting the Thickness of the Detector*

This standard technique is very commonly used in low- and high-energy radiation detection.

(1) Detection of x rays in the presence of gamma rays. Thin sodium iodide crystals with beryllium windows are used for the detection of x rays. The number of higher energy gamma rays stopping in these crystals is very small. A proportional counter made of low-Z material, such as aluminum with a beryllium window, is a good x-ray detector and its efficiency for gamma rays is very low. The total amount of material in the counter is not enough to stop the high-energy gamma rays, while they effectively stop x rays. The low-energy cutoff in both cases is determined by the thickness of the beryllium window.

(2) Detection of alpha particles in the presence of electrons. Detectors of thickness equal to the range of the particles to be detected are used in situations such as this one. Alpha particles produce signals much greater than the signals produced by electrons, which pass through the crystals with only very little loss in energy. Since the signals are of different height, pulse height selection can be easily used to count alpha particles only.

4.15.2. *Particle Selection by Specific Energy Loss*

In this method two detectors, one thin and the other thick, are used. The thin detector gives a signal proportional to dE, and the thick detector gives a signal proportional to $E' = E - dE$. As we have seen in Section 2.4, the product of dE/dx and E' is proportional to the mass of the particle and to the square of the charge of the particle. An approximate equation for this product P is

$$P \propto M^{0.8}Z^2E^{0.2} \tag{4.7}$$

Note that the product is comparatively insensitive to E.

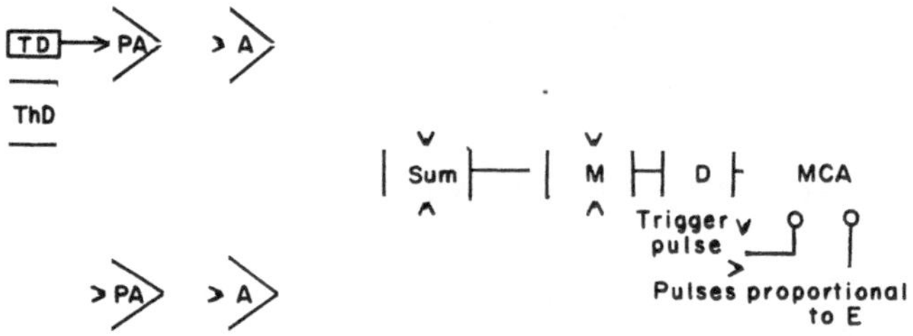

Figure 4-26. Block diagram illustrating the principle of particle selection by specific ionization. TD, thin crystal; ThD, thick crystal; PA, preamplifier; A, amplifier; Ad, adder; M, multiplier; D, discriminator; MCA, multichannel analyzer.

A block diagram of the experimental setup is shown in Fig. 4.26. The signals are added in the summing circuit, which produce a pulse proportional to the total energy of the radiation. The pulse from the multiplier is fed into a single-channel analyzer, where the signal corresponding to the desired radiation is selected. These signals are fed into the multichannel analyzer and they select the pulses from the summing circuit to be analyzed by the multichannel analyzer. The pulse height distribution seen on the display of the MCA will therefore be of the radiations selected by the single-channel analyzer. The thick detector is usually NaI(Tl) or CsI(Tl). Proportional counters, thin plastic scintillants, and totally depleted Si detectors have been used for the thin detector.

4.15.3. *Particle Identification by Emission Spectrum*

It has been observed that the wavelength distribution of the optical emission spectrum from CsI(Tl) depends on the incident particle. The relative intensity of light in the range 410–500 nm decreases with increase in the mass number. By using optical filters and comparing the intensities of light in the yellow and blue regions, it has been possible to distinguish alpha particles in the presence of gamma rays.

REFERENCES

1. W. J. Van Sciver, High Energy Physics Lab. Rept., No. 38, Stanford University (1955).
2. L. Pauling and E. B. Wilson, Jr., *Introduction to Quantum Mechanics*, McGraw-Hill, New York, 1955, pp. 309–311.
3. J. H. Neiler and P. R. Bell, The scintillation method, *Alpha- Beta- and Gamma-Ray Spectroscopy*, (K. Seigbahn, ed.), North-Holland Publishing Co., Amsterdam, 1965, Chapter V.
4. J. B. Birks *Theory and Practice of Scintillation Counting*, Pergamon Press, London, 1964.
5. J. H. Neiler and W. M. Good, in *Fast Neutron Physics* (J. B. Manion and J. L. Fowler, eds.), Interscience, New York, 1960, p. 509.
6. P. J. Ouseph, *Am. J. Phys.* **41**, 589 (1973).
7. F. D. Brooks, in *Progress in Nuclear Physics* (O. R. Frisch, ed.), Vol. 5 Pergamon Press, 1956, p. 252.

BIBLIOGRAPHY

J. B. Birks, *Theory and Practice of Scintillation Counting*, Pergamon Press, London, 1964.

J. H. Neiler and P. R. Bell, in *Alpha-, Beta-, and Gamma-Ray Spectroscopy* (K. Seigbahn, ed.), North-Holland Publishing Co., Amsterdam, 1965, Chapter V.

5

Semiconductor Detectors

The operation of a semiconductor detector is analogous to the operation of an ionization chamber. In an ionization chamber the incident radiation produces positive ions and electrons and an electrical signal is obtained by collecting these ions. In a semiconductor counter the incident radiation produces electrons and holes and information about the radiation is obtained by collecting them. One major difference of course is that only 3.5 eV has to be expended to produce an electron–hole pair in a semiconductor, while the value is ~30 eV in an ionization chamber. The lower energy increases the number of electron–hole pairs per MeV of radiation and decreases the relative statistical variation in this number. Hence the energy resolution is increased, that is, R is decreased.

5.1. OPERATION OF A SEMICONDUCTOR COUNTER

To understand the operation of the semiconductor counter, consider a bar of a uniform crystal with electrodes attached at

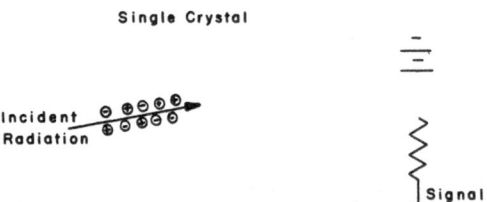

Figure 5-1.   Single-crystal semiconductor counter.

both ends (Fig. 5.1). When radiation falls on it, electron–hole pairs are produced, and when they are collected by the electrodes, a signal is obtained across the resistance, the height of which is proportional to the energy expended by the radiation in the material. The number of electrons produced by the radiation should be much larger than the number of electrons already in the conduction band. In other words, the steady state current should be very small. This will ensure that variations in the current will not be mistaken for a radiation-induced pulse. Therefore a good material for such a counter should be of low conductivity, preferably a perfect insulator at the operating temperature. This requirement eliminates metals and we are left with semiconductors and perfect insulators.

Energy resolution is usually an important factor in selecting a detector material. Energy resolution depends on the number of electron–hole pairs produced by the radiation. The energy required to produce an electron–hole pair is lower in a semiconductor than in an insulator. Hence the resolution of semiconductor detectors will be much better than that of detectors made of insulators. This requirement, therefore, is contradictory to the requirement of low conductivity discussed above.

The electron–hole pairs produced should be free to move in the detector and should be able to reach the electrodes. If the detector contains traps produced by impurities and defects, the electrons may be captured before reaching the electrode. Therefore only single crystals with the least number of traps are used for detectors. Trapping centers enhance recombination of holes and electrons, which will also reduce the pulse height. For a detailed discussion of the above two factors, let us start with the energy level diagrams of a semiconductor and an insulator (Fig. 5.2). The energy gap, the width of the forbidden band, is 1.08 and 0.67 eV for Ge and Si, respectively. Due to the thermal energy, some of the electrons may be able to cross the energy gap. The number of electrons thus reaching the conduction band for pure (intrinsic) semiconductors is

$$n_i = AT^{3/2} \exp(-E_g/2kT) \qquad (5.1)$$

where A is a constant for a given material (2.8×10^{16} for Si and 9.7×10^{15} for Ge), T is the temperature, E_g is the band gap, and k is Boltzmann's constant. At room temperature ($T = 300°K$), the electron density estimated using the above equation for Si ($E_g = 1.08$ eV) is $\sim 10^{10}/cm^3$ and for Ge is $\sim 2 \times 10^{13}/cm^3$. The possible statistical variation in this number for Si, $(10^{10})^{\frac{1}{2}} = 10^5$, is about the same as the number of electrons produced by a 1-MeV radiation,

$$\frac{10^6 \text{ eV}}{3.5 \text{ eV/electron–hole pair}} \approx 3 \times 10^5$$

(For silicon, E_g, or the energy required to take an electron from the top of the valence band to the bottom of the conduction band,

Conduction band

Eg

Valence band

Figure 5-2.   Energy level diagram of electrons in
a semiconductor (Si).

is 1.08 eV, but the average energy required for one electron–hole pair is 3.5 eV). Therefore it is difficult to distinguish pulses produced by radiations from those due to variation in electron density. From this discussion it is evident that a small energy gap is required for better energy resolution, whereas a large energy gap is required to reduce the steady state current and its fluctuations.

The electron in the conduction band is free to move and if an electric field is applied, it moves toward the positive electrode. The average speed, called the drift velocity, with which the electron moves under the influence of an electric field is a function of the strength of the field E, and the relationship between the two quantities is given by

$$v_e = E\mu_e \qquad (5.2)$$

where μ_e is called the mobility of the electron. There is a similar equation for the holes. Mobility for electrons and holes is a function of the temperature and the properties of the material. As mentioned, earlier, free electrons and holes may be trapped by the defects or impurity atoms before reaching the electrodes. The average time these charge carriers can move "freely" in the crystal before being trapped is called the lifetime. The product of mobility, lifetime, and the applied field is known as the "drift length," which should not be smaller than the size of the crystal for the collection of charge at the electrodes. From Table 5.1 it can be seen that the drift length for Ge is 1.5 cm²/V and for Si is 1.8 cm²/V. For a 1-cm-long crystal, an applied voltage of 1 V (field of 1 V/cm) will give a drift length of 1.5 and 1.8 cm for Ge and Si, respectively. One of the reasons for the failure of earlier "crystal counters" using small crystals of diamond, ZnS, and CdS developed by Van Heerden (1945) and Chynoweth (1952) was that the drift length was less than the size of the crystal.

From this discussion we can see that germanium and silicon will make good counter materials if the "free" charge density can be reduced. One method is to reduce the temperature of the material, which alone is not always enough because of the presence of impurities in the crystal. A second method is to use a

Table 5.1. Properties of Silicon, Germanium, and Diamond

	Silicon	Germanium	Diamond
Energy gap at 300°K, eV	1.08	0.67	6
Electron mobility at 300°K, cm²/V-sec	1500	3800	1800
Hole mobility at 300°K, cm²/V-sec	500	1800	1200
Electron lifetime in p-type, sec	3×10^{-3}	10^{-3}	—
Hole lifetime in n-type, sec	3×10^{-3}	10^{-3}	—
Atomic number	14	32	6
Mobility × lifetime, cm²/V	1.5	1.8	—

semiconductor *p–n* junction in reverse bias. McKay in 1949 used a point-contact junction to detect alpha particles. The diameter of the detector was only 10^{-3} cm. In 1956, Mayer and Gossick showed that gold-coated germanium, where a junction forms close to the surface, was sensitive to alpha particles. The advantage of this device was a large sensitive area. In a few years cooled germanium junctions were in wide use; later, similar silicon detectors were found to be more practical. In order to explain this operation of junction detectors, we shall first consider the semiconducting nature of silicon when impurities are added to it.

5.2. IMPURITY SEMICONDUCTORS

Figure 5.3 is a two-dimensional representation of a silicon crystal. Binding between the atoms is covalent and each atom contributes four valence electrons toward the covalent binding. If an atom like phosphorus, with five valence electrons, is substitute for a silicon atom in the crystal, one of the electrons will not participate in the binding of the crystal atoms. This electron can be regarded in first approximation as existing in a dielectric matrix of silicon and as forming a hydrogenlike atom with the net positive charge of the phosphorus atom. On this simplified model, since the relative dielectric constant of silicon is 12, the binding

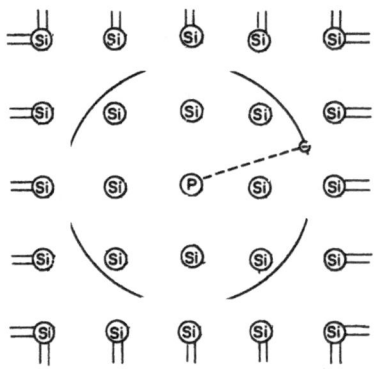

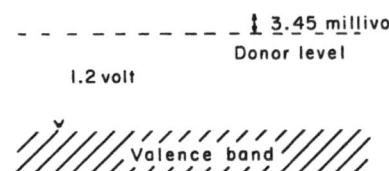

Conduction band

Donor level

1.2 volt

Valence band

Figure 5-3. Two-dimensional representation of silicon crystal containing phosphorus impurity. Position of impurity level is 3.45 mV below the conduction band.

energy is $13.6/12^2 = 0.090$ eV.* A more rigorous treatment gives a lower value, 0.045 eV. Thus the electrons of the impurity atom can be regarded as located at an energy level 0.045 eV below the bottom of the conduction band. Usually at room temperature there is sufficient lattice vibrational energy to excite this electron

---

* The equation for the energy levels of a hydrogenlike atom with one $+e$ charge at the center is $E_n = (me^4/8K^2\epsilon_0^2h^2)(1/n^2)$, where m is the mass of the electron, e is the charge, K is the dielectric constant, ϵ_0 is the permittivity, and h is Planck's constant. Substituting the values of the constants, we get $E_n = -(13.6/K^2n^2)$ eV. This means that, to ionize the atom, to release the electron from a hydrogen atom, an energy of 13.6 eV from the lowest state should be supplied to the hydrogen atom. For the hydrogenlike system in silicon, the electron–positive-ion distance is much higher than the corresponding distance in the hydrogen atom. Therefore K in the above equation is no longer equal to 1, but it is the dielectric constant of the medium, which is 12 for Si.

into the conduction band. Thus each substitutional phosphorus atom can contribute an electron to this band. A silicon crystal doped with phosphorus atoms is an *n*-type semiconductor.

On the other hand, if an atom such as indium atom, with only three valence electrons, is substituted, one of the four pair bonds leading from the indium atom to the four germanium atoms is incomplete. The indium atom is called an acceptor atom since it can take up an electron from the valence band, leaving a vacancy in this band. The impurity level in this case is located 0.16 eV above the top of the valence band. A silicon crystal doped with indium atoms is a *p*-type semiconductor. The vacancy left in the valence band when an electron is excited into the acceptor impurity level is called a "hole." Under the action of an electric field, an electron can move into the vacancy and thus shift the vacancy in the direction of the field.

Of particular interest in solid state detectors is the *p–n* junction. The juxtaposition of a region of *n*-type semiconducting material and a region of *p*-type semiconducting material forms such a junction at the interface. Its value lies in its rectification property, which we shall now describe qualitatively.

In thermal equilibrium the conduction electrons contributed by the donor atoms (the impurity atoms in an *n*-type semiconductor) predominate in the *n* region. Here the conduction electrons neutralize the space charge of the donor atoms. Similarly, the holes contributed by the acceptor atoms (the impurity atoms in the *p*-type semiconductor) neutralize the space charge of the acceptor atoms in the *p* region. In addition, at the junction between the two regions, diffusion plays a role in that electrons move into the *p* region and holes into the *n* region. In this process the electrons leave behind positively charged donor ions and the holes leave negatively charged acceptor ions. These ions are fixed and form a layer of opposite charges. The electric field of this layer prevents further diffusion across the junction.

However, the charge layer is not completely stable. A small flow of holes takes place from the *p* region to the *n* region, the holes eventually recombining with electrons. This loss of holes from the *p* region is compensated, as a consequence of thermal

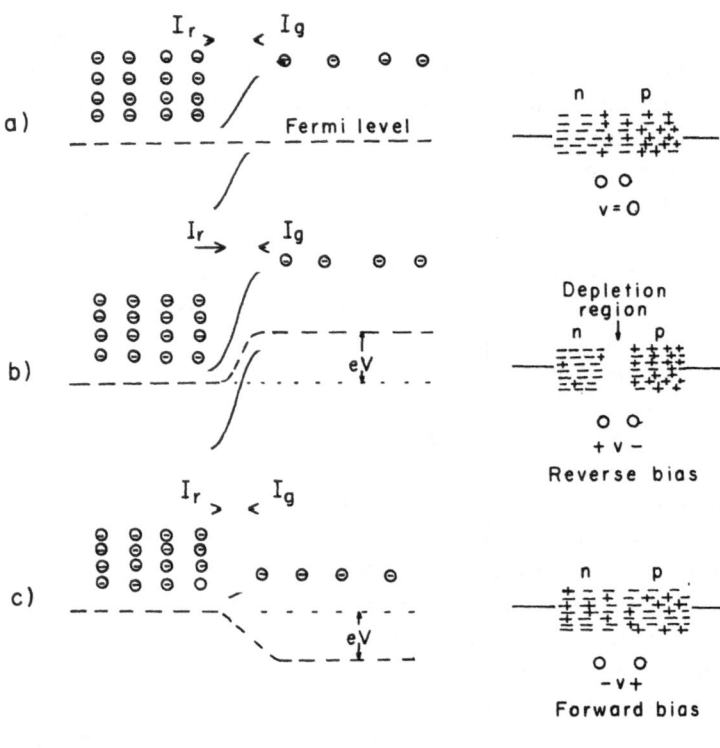

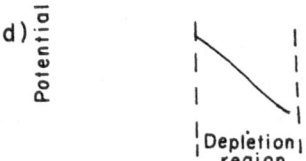

Figure 5-4. Energy level diagrams for the electrons in the p–n junction. (a) Normal p–n junction. Energy bands in both p and n semiconductors adjust so that there is no net current flow through the junction. The bottom of the conduction band on the n side is lower than the p side. The electrons on the n side therefore see a potential barrier to overcome to go to the p side. On the other hand, electrons from the p side are free to move to the n side. The number of electrons on the n side that can cross to the p side, i.e., the number with energy above the barrier, is equal to the number of electrons on the p side, making the currents in both directions equal. (b) Effect of reverse bias on the

fluctuations, by the generation in the n region of holes which diffuse into the p region. At equilibrium the recombination current i_r and the thermal generation current i_g balance, as indicated in Fig. 5.4(a). Recombination and compensation of electrons can be explained on a similar basis.

As a consequence of this behavior, the application of a potential difference across the junction will produce rectification. For reverse voltage bias, the p region is negative relative to the n region, as shown in Fig. 5.4(b), and the potential difference across the junction is increased. This makes it more difficult for holes to flow into the n region. Thus i_r decreases while i_g remains about the same. For forward bias, the p region is positive relative to the n region, as shown in Fig. 5.4(c), and the potential difference decreases. In this case i_r increases according to the strength of

p–n junction. The positive electric potential on the n side reduces the potential energy of the electrons, while the negative potential on the p side increases the potential energy of the electrons. The potential energy difference increases by the product of electronic charge and the applied voltage (eV). The number of electrons flowing from the n side to the p side decreases because of the increase in the height of the potential barrier. Therefore current from the n side decreases with voltage while I_g, the current from the p side, stays constant. As the reverse bias is increased the current (due to the few electrons on the p side) attains a final small value (see Fig. 5-5). Another result of the application of the reverse bias is the removal of charges from the junction of the diode. Positive potential attracts and collects the electrons to the n side and the negative potential attracts and collects the holes to the p side. The net effect is an almost charge-free volume at the center. This volume is called the depletion region. The depletion region free of charge carriers has the properties of an intrinsic semiconductor. (c) When forward bias is applied, the difference in potential energy decreases, making it easier for the large number of electrons on the n side to flow to the p side, resulting in an increase in current. (see Fig. 5.5.) Also note that the position of the Fermi level with respect to the bottom of the conduction band and the top of the valence band does not change in the two regions (dashed lines). The difference between the Fermi levels in the two regions gives the difference in energy (eV) of the electrons in the two regions. (d) The approximate electric potential distribution in the junction during reverse bias. Because of the concentration of charges at the ends, these regions behave as conductors and therefore there is no change in potential. The depletion region is of high resistance (no charge carriers). The potential drop is confined to this region. An electric field exists mainly in the depletion region. Charges produced in this region are swept to the electrodes by the field. Therefore only particles falling in this region will produce an electric signal and will be detected.

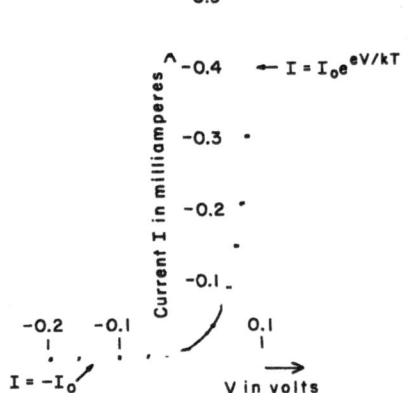

Figure 5-5. Variation of current with voltage in a *p-n* junction.

the field. Again i_g remains about the same. The electron currents across the junction behave similarly. The net current flow across the junction as a function of the applied voltage is shown in Fig. 5.5.

The application of the *p–n* junction to semiconductor detecttors depends upon the characteristics of the junction under reverse bias. As discussed in Section 5.1, one of the requirements for a good detector is reduced current flow in order to decrease current fluctuations. In reverse bias of the *p–n* junction, this condition is fulfilled without changing the energy gap E_g and therefore the energy resolution. Because of the absence of electrons and holes in the neighborhood of the interface, as indicated in Fig. 5.4(b), the sensitive volume of the detector is limited to this region, whereas in crystal counters the entire volume is sensitive. In reverse bias, the holes accumulate closer to the negative electrode and the electrons closer to the positive electrode. The region in the middle is practically free of charge and is therefore called the depletion region. The potential drop across the detector is essentially confined to the depletion region, as shown in Fig. 5.4(d). Consequently, if radiation enters the depletion region, electron–hole pairs produced in this region will be collected by the electrodes. If radiation falls outside this region, the electron–hole pairs, which are subjected to practically no potential difference, will rarely be collected at the electrodes.

The width of the sensitive volume (depletion region) depends on the applied bias voltage V and the resistivity ρ of the sample. For a p-type silicon diffused junction detector, the width W is approximately given by

$$w = [\rho(V + V_0) \times 1.1 \times 10^{-9}]^{\frac{1}{2}}$$

where V_0 is the potential barrier across the p–n junction at zero bias voltage.* The voltages V and V_0 are expressed in volts, the resistivity ρ in ohm-cm, and the width W in centimeters. The voltage V_0 is of the order of 0.3 V for silicon. The ultimate maximum in width depends upon the effective purity of the material and upon the breakdown voltage of the junction. For example, a p-type silicon base material of 10,000 ohm-cm resistivity has about 4×10^{12} acceptors/cm^3. At a reverse bias of 500 V, a width of 0.7 mm results. Another method of increasing the width, by a lithium diffusion technique due to Pell, is discussed later.

The reverse biased detector has three regions: the depleted region of high resistance and on either side of it a highly conducting layer where the charge carriers have accumulated. Hence it is similar to a plane capacitor. The capacitance of the system therefore depends on the parameters of the sensitive volume in the following way:

$$C = kA/4\pi W \tag{5.3}$$

where A is its area, W is the width, and k is the dielectric constant. Lower capacitance produces an increase in signal and also an increase in signal-to-noise ratio. Further discussion on the significance of the capacitance is given in the section on preamplifiers. The capacitance of a junction in the above example is about 10 pF/cm^2.

* For an n-type silicon surface barrier junction detector the constant in the equation 5-2 is 3.2×10^{-9}.

The fact that E_g is very small accounts for the superiority of semiconductor counters over other types of counters. A radiation of 1 MeV energy will produce $10^6/3.5 = 2.8 \times 10^5$ electron–hole pairs in silicon, while it produces about $10^6/35 = 2.8 \times 10^4$ electron–ion pairs in a gas counter operated at low voltages. It may be noted that the pulse height produced by these electron–hole pairs in a 1-cm² silicon junction counter of the previous example is only 4.5 mV:

$$\text{pulse height} = \frac{2.8 \times 10^5 \text{ charges} \times 1.6 \times 10^{-19} \text{ C/charge}}{10 \times 10^{-12} \text{ F}} = 4.5 \text{ mV}$$

The statistical variation in the number of electron–hole pairs produced is given by $n^{\frac{1}{2}}$, where n is the number of electron–hole pairs. Because of this, not all the pulses produced by radiation of the same energy will have the same height. This gives a finite width to the curve representing pulse height versus number of pulses. The full width of this curve at half maximum height (FWHM) expressed in terms of the pulse height is given by

$$\Delta E/E = 2.35 F^{\frac{1}{2}} n^{\frac{1}{2}}/n \tag{5.4}$$

The system will have better resolution if the FWHM is smaller. From the above examples it follows that the FWHM is 0.19% for semiconductor counters and ~0.7% for ionization counters. The Fano factor F is taken to be 0.2 for Si and 0.4 for gas counters. Other factors contributing to ΔE are discussed in Section 5.5.

5.3. DETECTOR TYPES

Reverse biased *p–n* or *n–p* junctions are used as nuclear detectors. A schematic diagram of such a detector is shown in Fig. 5.6. The *n* layer is made extremely thin, typically 0.1 μm, so that the energy loss in this layer is very small. Most detectors currently in use are of three different types, which are discussed below.

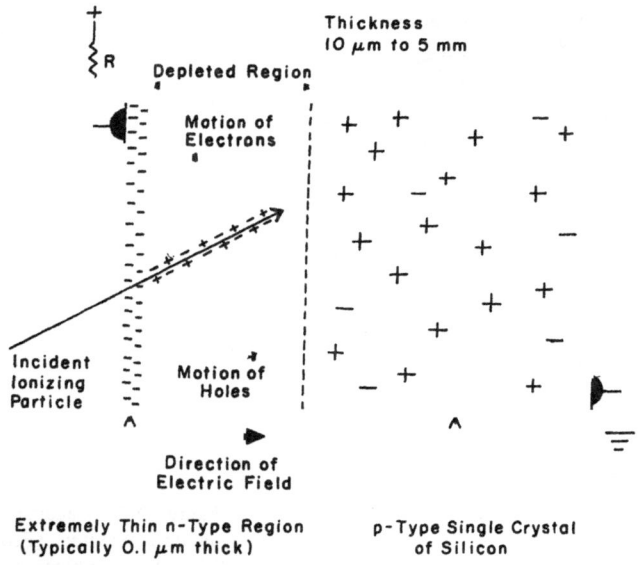

Figure 5-6. Schematic diagram of a *p–n* junction counter.

5.3.1. *Diffused Junction Detector*

This type of detector is usually produced by diffusing a high concentration of donor impurities into a *p*-type material. Silicon is usually used as the base material. Single crystals of high resistance are sliced into small, 1-mm pieces. Phosphorus is diffused into one surface of these slices. One commonly used method is to coat one side with phosphorus pentoxide dissolved in glycol and to heat the slice at 800°C in dry nitrogen for $\frac{1}{2}$ hr. The phosphorus diffuses into the base material while the glycol leaves the base material covered with a black residual deposit. After diffusion, proper electrical connections are made to the *p* and *n* sides. A typical counter arrangement is shown in Fig. 5.7. Special care is taken to reduce surface current and retard deterioration of the crystal. The metal container is filled with nitrogen or sometimes maintained under vacuum. One should be very careful in handling the detector, especially its surface.

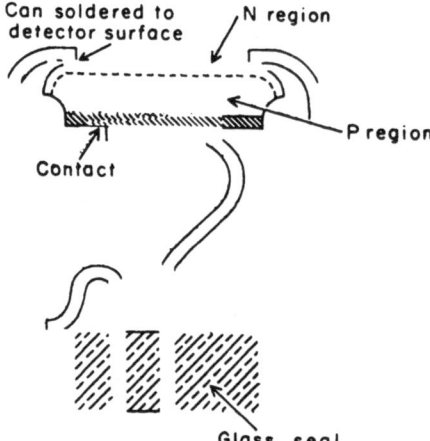

Figure 5-7. Encapsulated *p–n* junction detector. [From: R. L. Williams, "Semiconductor Particle Detectors," Nat. Acad. Sci. (Washington, D.C.), Publ. 871 (1961).]

Diffusion counters have been prepared by diffusing acceptors such as boron and gallium into *n*-type silicon crystals.

5.3.2. Barrier Layer Detectors

If *n*-type silicon is exposed to air, the surface layer oxidizes. A thin gold coating is applied to this surface. The layer close to this coating has the characteristics of a *p* layer. Gold plays an important role in the formation of the *p* layer. Other metals (except Al), when coated on similarly treated Si, do not produce a rectifying junction. The junction formed this way is closer to the surface. Hence energy losses by the radiation outside the active volume are minimal. Connection to the back side is through a nonrectifying metal contact. A typical counter arrangement is shown in Fig. 5.8.

Gold-coated surface barrier detectors are very delicate. Special care should be taken in handling them. Some of the precautions are; (1) Never touch the gold surface. (2) Since the gold layer is very thin, chemical vapors can diffuse through this surface. Therefore one should be mindful of the environment when the detector is being used. (3) Shielding against light should be provided when the detector is in use, since it is sensitive to photons. Other general precautions are to avoid mechanical and electrical shocks.

Recently ORTEC Inc. has developed a more rugged surface barrier detector made of *p*-type silicon. Aluminum coating on the *p*-type material produces the rectifying junction. The aluminum entrance window can be touched, cleaned, or decontaminated. This detector is light-tight and can be used in environments containing water vapor, dust, or light. ORTEC calls it a ruggedized silicon surface barrier detector.

Surface barrier Si detectors can be operated at room temperature. Similarly, germanium surface barrier detectors can be made. Since the energy gap is smaller, the counter cannot be used at room temperature. The energy resolution of germanium counters at 77°K is almost equal to that of silicon detectors at room temperature.

Equations (5.2) and (5.3) show the relationship between resistivity, bias voltage, depletion range, and capacitance for Si detectors. The relationships can be obtained from the nomogram

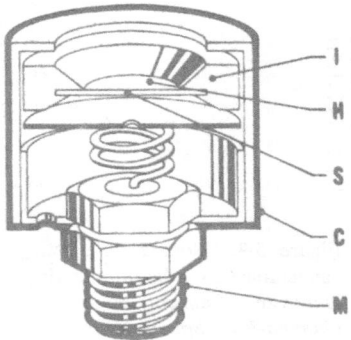

Figure 5-8. A typical barrier layer detector. [From: ORTEC Publication "Instruction Manual, Surface Barrier Detectors."] S, Silicon crystal; H, gold coating; I, ceramic ring with metallic coating on the front and back surfaces; C, metallic case; M, microdot connector.

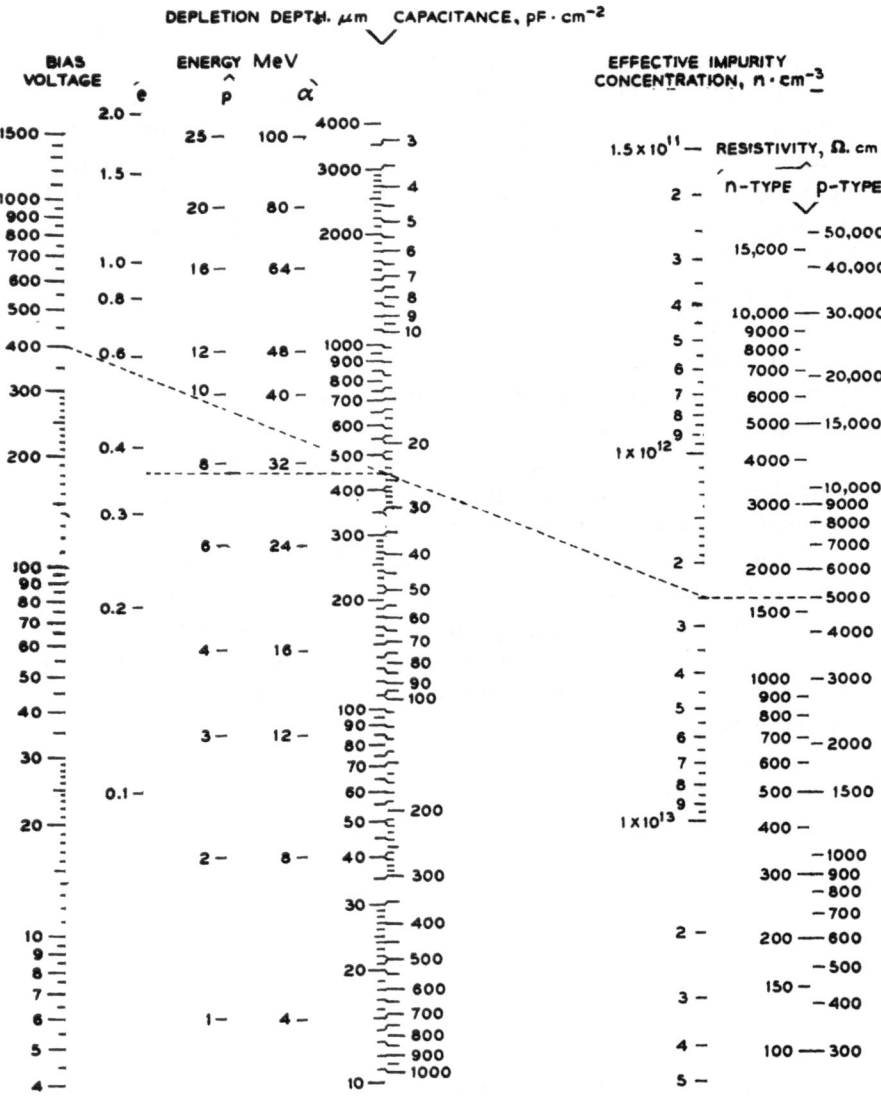

Figure 5-9. Nomogram relating resistivity, bias voltage, depletion range, and capacitance for silicon detecting. [From: W. W. Gibson, G. L. Miller, and P. F. Donovan, Semiconductor particle spectrometers, in *Alpha-, Beta-, and Gamma-Ray Spectroscopy* (K. Seigbahn, ed.), North-Holland Publishing Co. Amsterdam, 1965.]

in Fig. 5.9. Ranges of radiations for various energies are also shown in the nomogram. It is essential to determine the bias voltage so that the depletion width is slightly larger than the range of the particles under study. The resistivity of the detector is usually supplied by the manufacturer. Using this information, one can estimate the bias voltage from the nomogram. For example, let us assume one is interested in detecting 30-MeV alpha particles. Draw a horizontal line to the depletion width from the 30-MeV point on the alpha line. The reading at the point where the horizontal line meets the depletion width is the required value, 450 μm. This procedure also gives the capacitance per unit area for the 450-μm depletion width. It is interesting to note that for this depletion width, the energy of protons that can be detected is ~7.8 MeV and the energy of electrons is only ~0.35 MeV. To obtain the bias voltage, connect the point for the depletion width with the impurity concentration for the given resistivity of the detector. The value at the point where this line meets the bias voltage line gives the operating bias voltage. In our example this is 400 V.

5.3.3. *Lithium-Drifted Detectors*

Lithium is a donor atom. It does not go into substitutional sites like other donor atoms, such as phosphorus. Instead, it enters interstitial sites. The diffusion coefficient is about 10^7 higher than that for phosphorus and therefore deep diffused junctions can be prepared. Diffusion is usually achieved in two steps. First lithium is coated onto single crystals into which it is diffused by heating. In the second step the sample is heated and a strong reverse bias is applied. This helps the lithium atoms to diffuse deeper into the crystal. To a first approximation one obtains an ion distribution such as shown in Fig. 5.10. There are three regions: the *p* region, the *n* region, and an "intrinsic" region. The term "intrinsic" is usually used for crystals for which the free charge carriers of the material, and not the impurity-donated charge carriers, decide the electrical properties. However, in this case the last region results

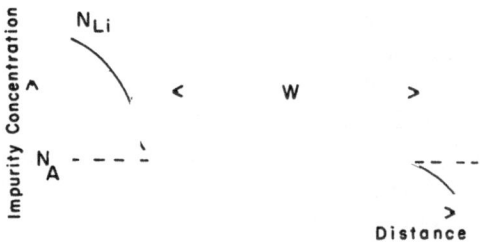

Figure 5-10. Lithium concentration as a function of distance from the face of the counter. In a region of width W, Li concentration N_{Li} balances the acceptor concentration N_A.

from the compensation or neutralization of the p-type impurities by the electrons donated by the lithium atoms. Even though it is an unfortunate choice, the term "intrinsic" is widely used in detector terminology. The intrinsic region constitutes the active volume of the detector. Active volumes of 50 cm³ have been achieved. This type of drifted detector is also known as a $p-i-n$ device. Lithium-drifted devices can be prepared in both germanium and silicon crystals.

The increase in the width of sensitive regions and the fact that germanium has a higher photoelectric cross section than silicon make lithium-drifted germanium detectors suitable for the

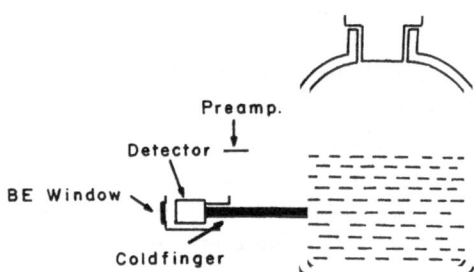

Figure 5-11. A typical liquid nitrogen crystal configuration for a Li-drifted detector. The field-effect transistor is placed on the cold finger. This reduces the noise.

detection of gamma rays. However, this advantage is slightly offset by the fact that the counter should be operated at 78°K because of the low energy gap and that the counter should be kept at temperatures below 150°K even when not in use, because the lithium compensation is not stable at room temperature.

Figure 5.11 shows a typical liquid nitrogen cryostat–detector–preamplifier assembly. Some times the field effect transistor of the preamplifier is placed on the cold finger close to the detector. Low temperatures and a short distance between the detector and the input stage of the preamplifier reduce the noise. The Ge(Li) detectors are available in planar or cylindrical (coaxial) shape.

5.3.4. *Intrinsic Germanium Detectors*

In lithium-drifted detectors, a sufficiently large active volume is achieved by compensating for the p-type impurities with electron-donating Li atoms. Electrons freed by the lithium atoms are picked up by the holes. Both types of impurities remain ionized; holes and electrons stay together without participating in the conduction. Thus the lithium-compensated region becomes a region of high resistance. The mobility of Li in the crystals at room temperature is one of the problems with these detectors. To prevent Li from drifting back, these detectors should be operated and stored at low temperatures, usually liquid nitrogen temperatures. Also, lithium ions tend to precipitate at sites of radiation damage, reducing the useful lifetime of the detector. These difficulties have encouraged investigators to look into the possibility of growing crystals of ultrahigh purity which will not require Li compensation. These attempts have been successful.

Let us now consider the Ge crystal at 78°K. The electrical conduction at this temperature is mostly due to charge carriers donated by the impurities [See Eq. (5.1) for intrinsic charge carrier density.] Therefore for a good detector the impurity concentration should be of the order of one impurity atom per 10^{13} Ge atoms. It is interesting to note that a material with an impurity concentration of one part in 10^6 is considered by chemists as a high-purity material. Crystals with impurity concentrations of one part in 10^{13} are now available commercially. It may be pointed out that methods have been developed to detect such low concentrations of impurities. Detectors fabricated with such crystals of size 200 mm² × 7 mm, suitable for x-ray detection, have a 200-eV resolution in the 6-keV range. The advantages of detectors made of these crystals are obvious. The detector can be cycled to room temperature and back to liquid nitrogen temperature without damage. Difficulties involved in fabrication are much less than for Ge(Li) detectors.

5.4. PULSE SHAPE AND RISE TIME

The shape of the pulse produced in a semiconductor counter depends on the position where the hole–electron pair is produced. As mentioned earlier, the drift velocity of the electron is higher than that of the hole. Therefore the induced voltage due to electrons will have a greater slope than the induced voltage due to the hole. Let us consider the three cases shown in Fig. 5.12. Pair 1 will produce a pulse with fast rise time, due to the motion of the electrons alone. The rise time of the pulse produced in pair 2 will have two parts: a rapidly rising part due to the collection of electrons and a slowly rising part due to the holes. Pair 3 produces slowly rising pulses since the electrons do not contribute to the rise time. The height of the pulse is equal to the charge of the electron divided by the capacitance of the system. An actual pulse will be a composite of pulses of varying rise times between the extremes of pairs 1 and 3.

In silicon at room temperature the mobility of the hole is 500 cm²/V-sec and the mobility of the electron is 1500 cm²/V-sec. The mobility is constant up to a velocity of 10^6 cm/sec and decreases for higher velocities. For applied field strengths of the order of 10^3 V/cm and depletion layers of 1 mm thickness, the rise time is 10^{-7} sec. Smaller rise times can be achieved by increasing the voltage or decreasing the width of the depletion layer. However, the resistance and capacitance of the system limit the rise time usually in the range of 10–100 nsec.

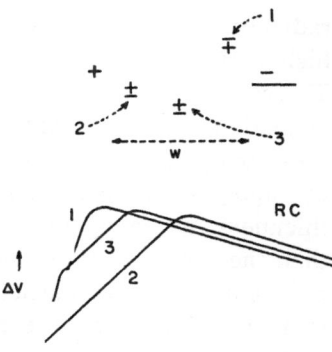

Figure 5-12. Pulse shapes due to electron–hole
pairs at different positions in the counter.

As indicated in Section 4.6, one of the chief advantages of semiconductor counters is the linearity between the energy of the radiation and the pulse height. In addition, this relationship is independent of the nature of the particle.

5.5. FACTORS AFFECTING ENERGY RESOLUTION

The outstanding feature of the semiconductor counter is its energy resolution, which is responsible for the superiority of this type of counter. The two basic factors determining the maximum possible resolution are the statistical variation in the number of electron–hole pairs produced by the radiation and the Fano factor. In deriving Eq. (5.4) for resolution, we considered these two factors only. There are other factors which contribute to the width of the peak and increase the value of R. These are discussed in this section, and include (1) the detector noise, (2) electronic noise, mainly that of the preamplifier, (3) imperfect charge collection, and (4) variation in the energy received by the active volume of the counter. The last two factors depend on the mass of the incident particle. The electric field in the crystal should be sufficient to collect the charge carriers produced by the radiation. The charge density produced by heavy particles is higher than the charge density produced by lighter particles. Therefore a higher electric field should be applied for complete collection of the charge produced by heavy particles. Line broadening due to incomplete charge collection can be controlled by applying proper bias voltage. The window (dead layer) thickness, assuming the thickness of the active volume is larger than the range of the particle, contributes to the line broadening, especially for heavy particles. For example, for alpha particles the minimum line width one can get appears to be 12–13 keV even with no significant contribution from noise. Also, for heavy particles energy losses in the source and air become signifcant. Alpha-particle spectroscopy experiments are therefore usually performed in vacuum using thin, uniform sources of small diameter.

The main source of detector noise comes from the fluctuations in the leakage current, both volume and surface currents. A defect or an imperfection may capture an electron and release it after a while. This process is known as carrier generation. If this is happening in the depleted region, the charges will be swept by the electric field, producing a signal. If the carrier generation happens in the undepleted region, the charges will first have to diffuse into the depleted region to produce a signal. The leakage current due to carrier generation in the depleted region is $<10^{-5}$ A/cm^3 and that due to carrier generation in the undepleted region is comparatively very small (10^{-8} A/cm^3). Surface leakage current, in addition to producing noise, also determines the breakdown voltage. Guard ring arrangements for the electrodes and proper chemical treatment help to reduce this current. Noise and other special preamplifier features are discussed below.

The preamplifiers play an important role in the total performance of a semiconductor counter system. Therefore a somewhat detailed discussion of them is given here. The preamplifiers in proportional and scintillation counting systems function mainly as a power amplifier, to supply enough power to transmit the signals along long cables to the main amplifiers. In these detectors, the internal amplification is high. Therefore preamplifiers do not have to amplify the signals, and since input signals are of sufficiently high amplitude, the noise produced by the preamplifiers is not a serious problem. However, there is no internal amplification in the semiconductor counters. Therefore here the preamplifiers should be capable of good amplification and should have extremely good noise characteristics. Preamplifiers with field effect transistors (FET) at the input stage satisfy the above requirements. Only FET amplifiers are now in use in counting systems.

The capacitance of the detector is a function of the bias voltage. The pulse height for a radiation producing a charge of Q is Q/C_D. Changes in C_D due to small changes of bias voltage will therefore affect the output pulse height. Line voltage changes can produce such small bias voltage variations. Also, bias voltage is often changed to adjust the depletion width according to the range of the particle being detected. These changes do not affect the total charge collected, which is always proportional to the energy of the radiation. Therefore charge-sensitive preamplifiers are used in semiconductor detector systems. The output from these amplifiers is proportional to the charge at the input. Voltage-sensitive and current-sensitive preamplifiers are used in gas and scintillation detector systems.

The equivalent circuit of a typical detector and the first stage of the preamplifier is shown in Fig. 5.13. Here C_D denotes the capacitance of the detector and Q is the charge produced by the radiation in the sensitive volume of the detector. The input to the preamplifier in this case is Q/C_D and the output voltage is approximately given by $-Q/C_f$, provided that $G_0C_f \gg C_t$, where G_0 is the open loop gain, C_f is the feedback capacitance and C_t is the

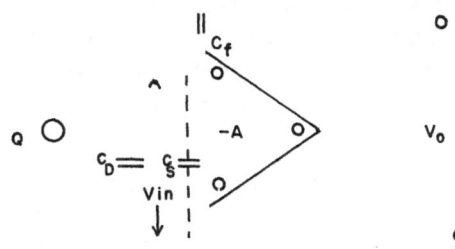

Figure 5-13. Equivalent electrical circuit for a typical detector and the first stage of a preamplifier.

total input capacitance, equal to $C_D + C_s$, where C_s is sum of the stray capacitances due to connectors, cables, etc. The output of the preamplifier is proportional to the charge Q and it depends on the fixed C_f and not on C_D. The amplification of the preamplifier is C_f/C_t. Thus the effect of the detector capacitor on the output signal is reduced by choosing a charge-sensitive preamplifier. The detector capacitor still acts as a source of noise to the preamplifier. The energy of the signal is given by $Q^2/2C_t$. When C_t increases, the energy decreases, reducing signal-to-noise ratio.

The physical parameters of the detector can be changed to decrease the detector capacitance. However, such changes may affect the operation of the detector unfavorably. The capacitance is proportional to the area of the detector, but reducing the area also reduces the detection efficiency of the detector. Another parameter that is easily adjusted is the width of the depletion region. Increasing the width decreases the capacitance. Two unfavorable results of this are (1) the carrier-generated noise in the detector increases, and (2) the rise time of the pulse increases. Therefore when adjustment of the parameters is made, consideration should be given to detector noise, preamplifier noise, and experimental requirements such as detection efficiency, pulse rise time, and particle range.

There are methods by which better energy resolution can be achieved. Magnetic spectrometers give a resolution for beta rays of the order of 0.02% in the keV range and for alpha particles 0.6% in the 6-MeV region. Also, it is interesting to note that a variation in energy of one part in 10^{13} can be detected by the Mössbauer effect even though the precision with which the absolute energy can be measured with a crystal diffractometer is only about one part in 10^4. However, semiconductor counters are advantageous with regard to (1) efficiency, (2) the possibility of simultaneously detecting radiations of different energy, (3) high resolution, and (4) simplicity of the whole system.

5.6. *RADIATION DAMAGE*

A defect-free crystal is essential to obtain maximum performance from a detector. A charged high-energy particle passing through the crystal, in addition to producing electron–hole pairs, may produce localized defects. At times the particle deposits enough energy at the crystal lattice to displace an atom to an interstitial position. The vacancy thus created with the displaced atom is known as a Frenkel pair. This defect acts as a carrier trap, resulting in increased leakage current and consequently higher detector noise. Traps hinder the free flow of charge carriers, reducing resolution and sometimes resulting in additional peaks on the low-energy side of the main peaks. The severity of the deterioration of the performance depends on the total dosage and nature of the particle. The effect of radiation damage appears to be more serious in Li-drifted detectors. Table 5.2 summarizes the effects of different radiations on detectors.

The performance of gas counters and scintillation counters also decreases with the total radiation detected, although the effect is less pronounced than in semiconductor counters. In gas counters the only effect is the breaking up of the quenching gas, resulting in slow changes in operating characteristics. In scintillation counters, small decreases in luminescence efficiency have been noted.

5.7. DETECTION OF CHARGED PARTICLES

All types of detector—diffused junction, surface barrier, and Li-drifted—can be used for charged particle detection. Silicon crystals are usually used as a base material for charged particle detectors. Since the atomic number is low, backscattering of electrons will be very small. These detectors can be operated at room temperature. However, when good energy resolution is needed, low-temperature operation is preferred. Since the detectors have no window, their efficiency for charged particles is almost 100%, except when the particle energy is very low. For low-energy particles, the height of the pulses produced may be lower than the noise pulses. When the discriminator in the detector system is adjusted to eliminate the noise pulses, some of the low-energy electron pulses will be lost. Therefore the detection efficiency for low-energy beta rays, such as rays from

Table 5.2. Radiation Damage in Silicon Semiconductor Detectors[a]

Detector type	Incident radiation	Energy, MeV	Total dose in ref. experiment	Dose for significant deterioration	Dose for device failure	Type of failure
Diffused junction	^{60}Co gammas	~1.25	10^7 r	10^6 r	—	Increase in capacitance from 10^6 to 10^7; resolution and noise deteriorate badly during exposure, but recover almost immediately after exposure
Li drift	^{60}Co gammas	~1.25	10^7 r	10^6 r	2×10^7 r	Small but detectable changes in sensitivity and noise at 10^6 r; greatly decreased sensitivity and increased noise at 2×10^7 r
Li drift	^{60}Co gammas	~1.25	1.2×10^6 r	10^5 r	10^6 r	Collection efficiency decreased; pulse rise time increased; effects can be partially compensated by increasing bias voltage
Surface barrier	Electrons	2	2.3×10^{13} electrons/cm^2	No data	No data	Multiple peaking; some noise increase; multiple peaking removed by increasing bias voltage

Li drift	Electrons	1	15^{15} electrons/cm²	—	—	Reverse current very high during exposure; recovers rapidly after exposure; some increase in noise; no significant change after 1.6 $\times$ 10^{15} electrons/cm²
Surface barrier	Fast neutrons	Fission spectrum	10^{12} n/cm²	3.5×10^{11} n/cm²	2×10^{12} n/cm²	Resolution broadening; multiple peaking after 3.5 $\times$ 10^{11} n/cm²; no single peak response after 2 $\times$ 10^{12} n/cm²
Surface barrier	Neutrons	14	2.6×10^{12} n	3×10^{11} n	10^{12} n	Multiple peaking; resolution broadening, increased reverse current, decreased pulse height
Surface barrier	Alphas	5.5	10^{11} α/cm²	10^{8}–10^{9} α/cm²	10^{11} α/cm²	Increased reverse current and noise; resolution broadening; multiple peaking between 10^{9} and 10^{10} α/cm²

[a] Table adapted with permission from ORTEC publication "Instruction Manual for Surface Barrier Detector," p. 1–18.

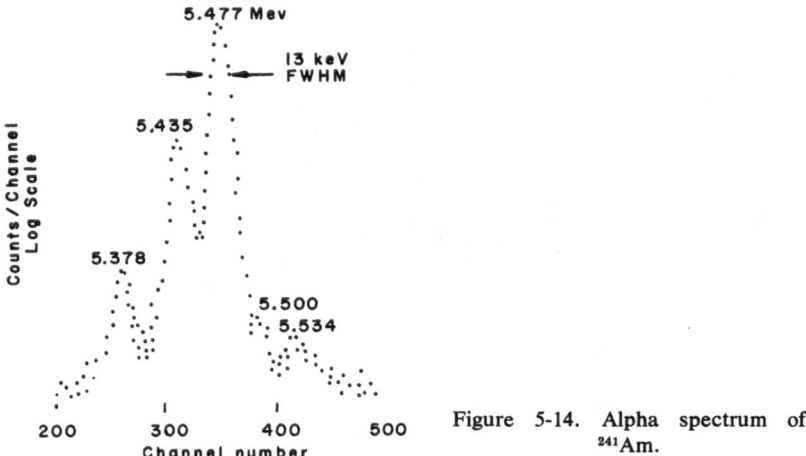

Figure 5-14. Alpha spectrum of ^{241}Am.

^{14}C sources, will be less than half of the efficiency of gas counters. However, the efficiency is higher than that of gas counters for high energy beta rays.

Semiconductor counters are also useful for alpha-particle spectroscopy. Generally, compared to magnetic spectrometers, semiconductor counters have poor energy resolution, but they offer better geometry. They can also be advantageous where magnetic spectrometers cannot be used because of low counting rate or short half-life of the source. Better energy resolution, lower background, better time resolution in coincidence work,

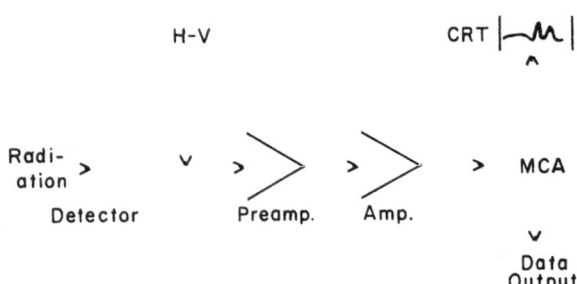

Figure 5-15. Block diagram of a typical spectroscopy system.

and higher permissible counting rate are some of the advantages of these counters over gridded gas counters. Figure 5.14 shows the alpha spectrum of ^{241}Am obtained with a surface barrier detector. The counter was operated at room temperature. The spectrum was obtained with an experimental setup similar to the one shown in Fig. 5.15.

For a good electron spectrometer system, the width of the depletion region should be of the order of the range of the electrons. Depletion layer thicknesses of the order of 0.8 mm (approximately the range of 0.6-MeV electrons) for diffused *p–n* junctions and 1.5 mm (the range of 1-MeV electrons) can be obtained for barrier junctions. However, greater depths can be achieved in Li-drifted silicon counters. Results of measurements by Ahmad and Wagner[1] with a cooled Si(Li) detector of size 80 mm^2 × 3 mm are shown in Fig. 5.16. For the 114.95-keV electron peak the FWHM is 880 eV (~0.77% resolution). in the 600-keV range the FWHM is ~2 keV. This high resolution was achieved

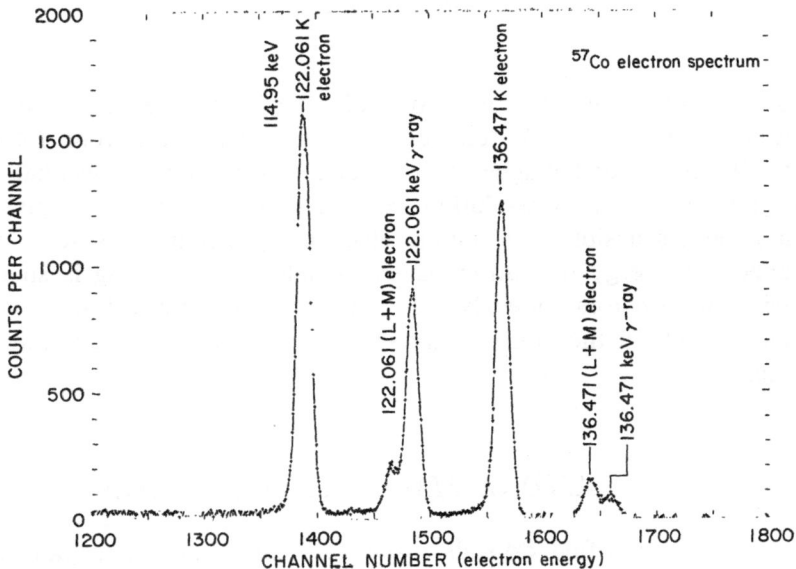

Figure 5-16. The ^{57}Co conversion electron spectrum measured with an 80 mm^2 × 3 mm, cooled Si(Li) spectrometer. [From: I. Ahmad and F. Wagner, *Nucl. Inst. Meth.***116**, 465 (1974).]

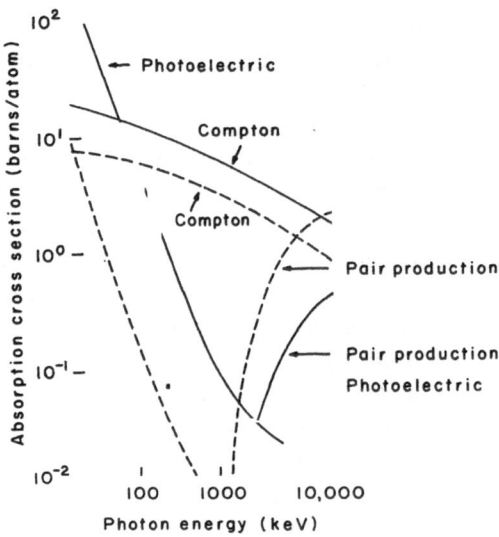

Figure 5-17. Photoelectric, Compton, and pair-
productive cross sections for silicon (dashed lines)
and germanium (solid lines).

by a commercially available low-noise FET amplifier and a low-
leakage-current Si(Li) detector. For low-energy electrons (<50
keV), Ahmad and Wagner found that the window thickness has a
significant effect on resolution and line shape. Their paper[1] gives
a good discussion of a simple electron spectrometer system of
excellent energy resolution. Even though backscattering is small
with Si detectors, one always gets a long tail toward the low-
energy side in the pulse height spectrum due to this backscattering.

5.8. X-RAY AND GAMMA-RAY DETECTION

Lithium-drifted crystals make very good x-ray and gamma-
ray spectrometers. Pulses are produced by electrons resulting
from any one or combinations of photoelectric, Compton effect,

and pair-production processes. Cross sections for these processes for Ge and Si are given in Fig. 5.17. From this figure it can be seen that for gamma rays the photoelectric cross section for Si is much smaller (about 100 times) than for Ge. Therefore silicon counters make very poor detectors for gamma rays compared to germanium detectors. However, they are excellent detectors of x rays in the presence of high-energy gamma rays. Figure 5.18 shows the variation of efficiency of Ge and Si counters with energy. The lowest energy limit is set by the counter window. On the high-energy side, for example, for a 62-keV x ray, the efficiency of the Si counter is 5% and that of the Ge counter is almost 100%. With cooled, commercially available, Li-drifted Si detectors a resolution of 198.8 eV for 614-keV x rays can be obtained.

From Fig. 5.17 it is clear that for 1-MeV gamma rays the

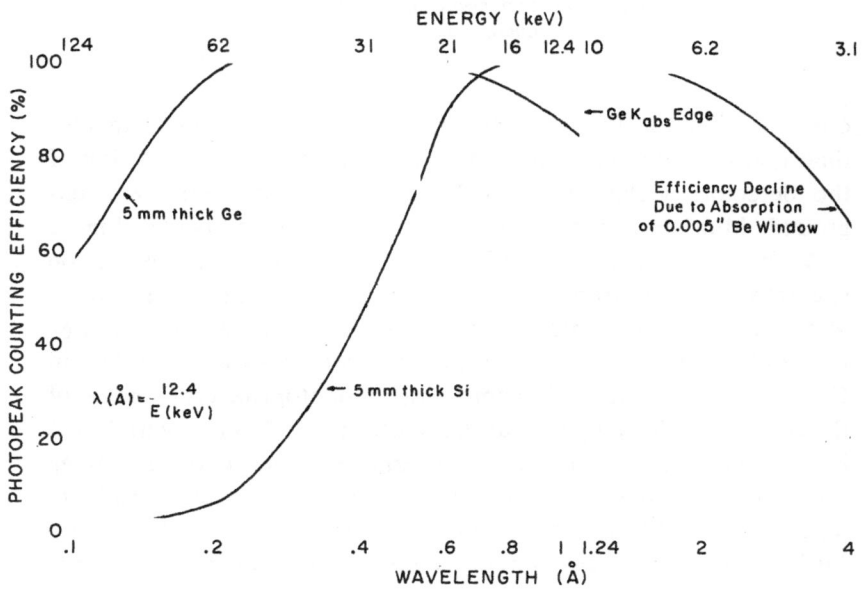

Figure 5-18. Variation of efficiency of Ge and Si counter with energy. [Data supplied by ORTEC Inc., Oak Ridge, Tennessee.]

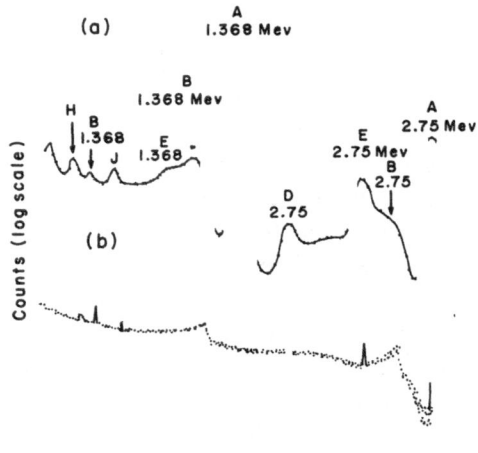

Pulse height

Figure 5-19. Comparison of ^{24}Na spectra obtained using NaI(Tl) and Ge(Li) detectors. ^{24}Na emits two gamma rays of 1.368 and 2.75 MeV. (a) The scintillation spectrum; (b) the semiconductor spectrum.

cross section for the photoelectric effect is very much smaller than that for Compton scattering. However, the high resolution of the detectors yields sharp peaks well resolved from the background. Multiple processes also contribute to these peaks. Figure 5.19 shows the spectra of ^{24}Na gamma rays from a scintillation spectrometer and from a Ge detector system. The tremendous improvement in resolution is very clear. Figure 5.19 also identifies the contributions of single and multiple processes as described in Chapter 4. For 1.33-MeV gamma rays, photopeak efficiencies of the order of 15% compared to the photopeak efficiency with 3 × 3 in. NaI(Tl) can be obtained. The large increase in the resolving power of Li-drifted Ge detectors counterbalances this disadvantage. A full width at half-maximum (FWHM) of 1.7 keV for 1.33-MeV gamma rays can be obtained in commercially available counters. Another advantage of Ge(Li) and other semiconductor counters is the fast rise time of the pulses.

5.9. NEUTRON DETECTORS

As in other types of detectors, semiconductor detectors of neutrons use neutron–proton elastic scattering, or one of the reactions ^{6}Li(n, γ)^{3}H and ^{3}He(n, p)^{3}H. In most neutron counting systems, a thin layer of the reacting material is kept between two surface barrier detectors so that the reaction products fall on these detectors. The sum of the pulses is proportional to the energy of the neutron plus the reaction energy. For detectors with ^{3}He at 5 atm of pressure, the resolution is 150 keV for thermal neutrons and the efficiency is 10^{-4}%. The energy resolution for counters using the Li reaction is about 300 keV for thermal neutrons and the efficiency is 0.1% for thermal neutrons and 10^{-4}% for 2.5-MeV neutrons.

5.10. OTHER APPLICATIONS

The tremendous improvement achieved in semiconductor counters has helped studies in several fields. The effect of this on nuclear spectroscopy is illustrated in Fig. 5.20. This figure shows how nuclear spectroscopic information improved first with the development of scintillation counters and then with the use of Ge(Li) detectors.[2] In most low- and intermediate-energy nuclear studies, such as those of charged particle reactions, nuclear fission products, and alpha-, beta-, and gamma-ray spectroscopy, semiconductor counters have replaced the other types of detector.

The development of these counters has improved the capability of x-ray fluorescence spectroscopy and activation analysis. These are widely used analytical methods for nondestructive testing. Fluorescence spectroscopy involves exciting the atomic inner electrons by x rays, radiations (α, β, or γ), or electron sources and analyzing and counting the x rays during deexcitation. The x rays emitted are characteristic of the atoms in the sample. A simple Si(Li) counting system can be used for this purpose. The positions of the x-ray peaks identify the atoms,

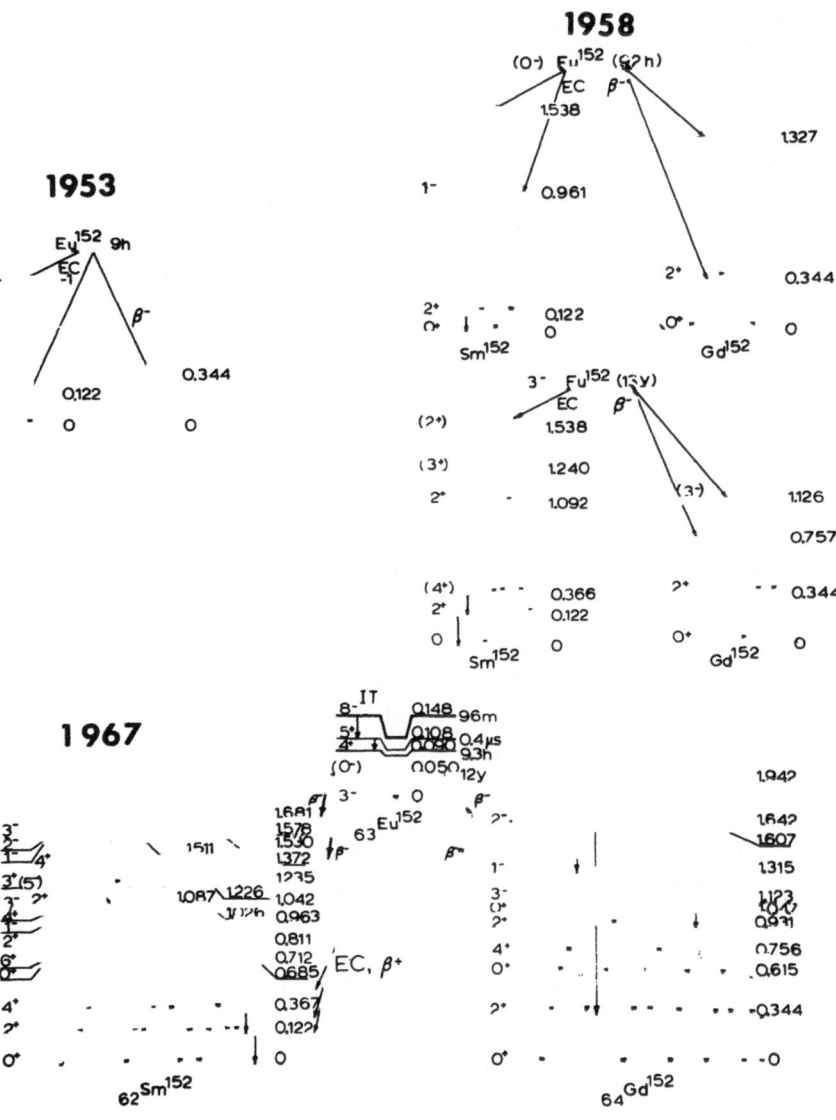

Figure 5-20.   The decay schemes of $^{152}_{63}$Eu. Data from the 1953, 1958, and 1967 Tables of Isotopes. This illuminating comparison, showing the progress made in nuclear spectroscopic studies with advances in detector technology, first appeared in an article by Hamilton *et al.*[2] The 1967 data were obtained with a Ge(Li) detector.

while the intensities can be used to estimate the amounts of different atoms in the sample. Almost all atoms except those with atomic number below eight can be analyzed by this method. As an example, let us consider chromium, manganese, and iron. The energies of the K x rays of these atoms are 5.4, 5.89, and 6.4 keV, respectively, so that the energy difference is 490 eV between chromium and manganese and 510 eV between manganese and iron. Energy resolutions of the order of $\sim$150 eV in the energy range of 6 keV are easily obtainable from commercial systems. Figure 5.21 shows the separated peaks due to these atoms. It also shows the difference in the performance of semiconductor and gas counters with respect to resolution.

Activation analysis involves identifying the elements in a sample from the radiations emitted by them after they are made radioactive. The elements are irradiated in a neutron flux or a

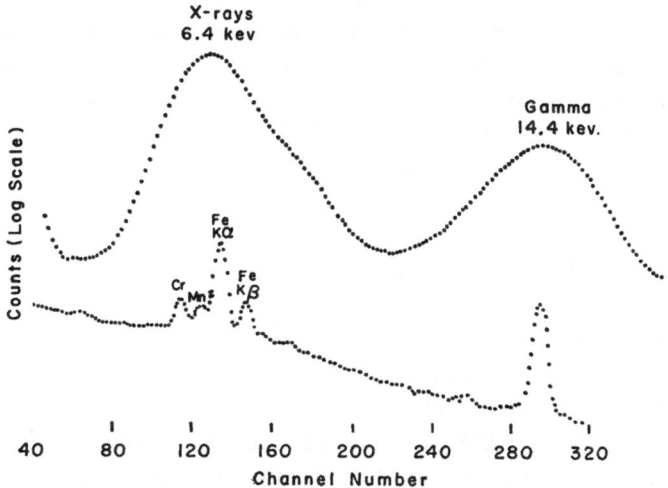

Figure 5-21. Pulse height distribution for 14.4-keV and 6.4-keV radiations emitted from a ^{57}Co source. The source is prepared by electroplating ^{57}Co on stainless steel. The ^{57}Co radiation excites chromium and manganese atoms in the steel, producing their characteristic xrays (5.4 and 5.89 keV). The top curve is the spectrum obtained with a proportional counter. Note the difference in resolution of the two types of detector.

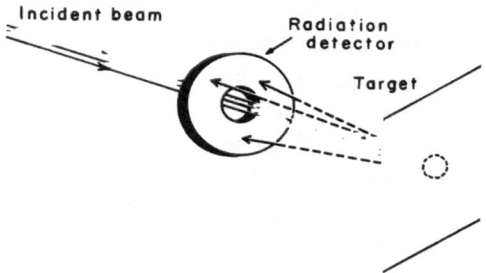

Figure 5-22. Experimental arrangement for studying backscattered particles by a ring counter.

charged particle beam. In some cases the reaction between these particles and the nuclei produce radioactive products. If the cross section of the reaction producing the radioactivity is known, the amounts of the elements can be estimated from the intensity of the radiations. The Ge(Li) counter, because of its high resolution, has the advantage of separating gamma rays of small energy differences. The bigger sizes of Ge(Li) detectors have also improved the efficiency of detection.

5.11. *SPECIAL TECHNIQUES*

Crystal counter detectors can be prepared in any shape. One of the interesting configurations is the annular type. This type permits the passage of charged particles through the detector to the target (Fig. 5.22) so that the particles backscattered, or reaction products going in the direction of the counter, can be detected.

Another type of counter available is the very thin, totally depleted counter. In such a counter the active volume extends from the front window to the back electrode. The thickness is made very small so that only part of the energy of the radiation is expended in this counter. The counter is placed on top of a second crystal, which has an active volume thick enough to stop the incident particle. We get two pulses from such an arrangement: one from the first counter, proportional to dE, the part of the energy expended in this counter; and the other from the second counter, proportional to E, the total energy. Using this information, one can identify the type of incident radiation. (See Section 4.15.2 for a typical experimental setup.)

Another type of detector, the nuclear triode, is capable of giving two separate signals, one proportional to the total energy of the radiation and the other proportional to the distance of impingement of the radiation from one end of the detector. Figure 5.23 is a schematic diagram illustrating the principle of such a detector. A surface barrier diode is formed on a long strip of *n*-type material with a conducting gold surface. On the back side is a resistive layer with two contacts, one grounded and the other connected to a charge-sensitive preamplifier. The charge flow through these contacts will be divided according to the resistance from the point of incidence to the respective contact. Therefore the pulse height E_x will have information about the distance between the point of incidence and terminal 3. The pulse height will be a maximum for zero distance. The charge collected at terminal 1 will be proportional to the total energy of the radiation. Counters with a position resolution of 1% of the total length for a 5-cm-long counter are available.

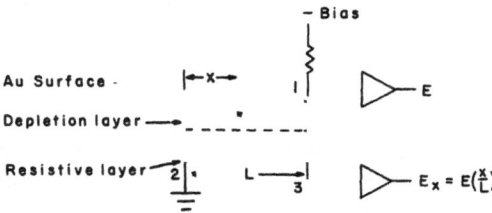

Figure 5-23. Schematic diagram of a position-sensitive nuclear triode.

5.12. *CONCLUDING REMARKS*

It is obvious that the most important advantage of semiconductor counters is the high resolution. The advangages and disadvantages as listed by Gibson, Miller, and Donovan[3] are given below.

Advantages
(a) Excellent energy resolution.
(b) Linear response over a wide range of particle types and energies.
(c) Insensitivity of pulse height to counting rate.
(d) Fast pulse rise time.
(e) Windowless operation.
(f) Variable sensitive thickness.
(g) Selectivity for charged particles.
(h) Small size and ease of handling.
(i) Insensitivity to magnetic fields.
(j) Ability to operate at low temperature.
(k) Ability to be fabricated into special structures, e.g., mosaics, annular devices, and position-sensitive detectors.

Disadvantages
(a) Inability to stop particles of long range.
(b) Relatively short operating lifetime of many currently available devices (due to surface degradation or radiation damage).
(c) Low output signal level.
(d) Relatively slow, and particle-range-dependent, signal rise time for thick devices.
(e) Inability to operate at high temperatures.

Almost daily, improvements are reported in the fabrication of detectors and the design of preamplifiers and other electronic components. Therefore by the time this book appears the values of resolution, efficiency, low-energy limit, etc. may be entirely different from those quoted here.

REFERENCES

1. I. Ahmad and F. Wagner, *Nucl. Inst. Meth.* **116**, 465 (1974).
2. J. H. Hamilton, J. C. Manthuruthil, and J. O. Rasmussen, *Nucl. Inst. Meth.* **113**, 277 (1972).
3. W. W. Gibson, G. L. Miller, and P. F. Donovan, Semiconductor particle spectrometers, in *Alpha-, Beta-, and Gamma-Ray Spectroscopy*, (K. Seigbahn, ed.), North-Holland Publishing Co., Amsterdam, 1965.

BIBLIOGRAPHY

See Ref. 3.

G. Dearnaley and D. C. Northrop, *Semiconductor Counters for Nuclear Radiations*. Wiley, New York, 1963.

J. M. Taylor, *Semiconductor Particle Detectors*, Butterworth and Co., London, 1963.

6

Corrections in Radiation Counting

6.1. *INTRODUCTION*

In general for any counting system, the number of counts registered by the scaler will be less than the number of radiations emitted by the radioactive source. For a simple counting arrangement, such as the one shown in Fig. 6.1, some obvious factors relating to this difference are the geometry of the system, the efficiency of the counter, absorption in the source, in air, and in the window of the counter, etc. These factors are discussed below.

6.1.1. *Geometrical Factor*

If the activity of the source is A, the number N of radiations reaching the counter (assuming no absorption and scattering) is

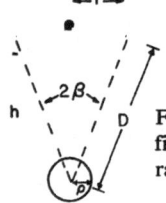

Figure 6-1. Simple experimental setup for radiation counting. The figure defines the angle β and the source has a finite radius ρ. The radius of the counter is r. The source-to-counter distance is h. $D = (h^2 + r^2)^{\frac{1}{2}}$.

given by

$$N = AG = A\Omega/4\pi \qquad (6.1)$$

where G is the correction factor for the geometry and Ω is the solid angle subtended by the window of the counter at the source. The geometrical factor G in Eq. (6.1) for a point source (radius $\rho = 0$) is (Fig. 6.1)

$$G_p = \Omega/4\pi = \tfrac{1}{2}(1 - \cos \beta) \qquad (6.2)$$

The subscript p indicates that the factor is for point sources. The angle β can be obtained from the equation $\tan \beta = r/h$, where r is the radius of the counter and h is the distance from the source to the window. In practice the source will have a finite radius. The geometrical correction factor for a finite size source is given by

$$G \simeq G_p\{1 - \tfrac{3}{8}\rho^2[h(h + D)/D^4]\} \qquad \text{for small } \rho/D \qquad (6.3)$$

where ρ is the radius of the source and $D = (h^2 + r^2)^{\frac{1}{2}}$.

From a series of very careful experiments, Zumwalt[1] has determined that G is independent of the energy of the beta particles. However, he also found that G as determined experimentally agrees with the theoretically calculated G only if h is taken to be about 4 mm more than the measured value. The reasons for the displacement of the sensitive volume to the back of the window of the counter are (1) termination of the center wire in back of the window, usually by a glass bead of finite size, and (2) accumulation of charge on the mica window. An experimental determination of G using a carefully prepared thin source deposited on a polystyrene–Formvar film (~ 50 $\mu g/cm^2$) will be better than an approximate determination of G by calculation.

6.1.2. *Self-Absorption and Scattering*

Absorption and scattering of radiation in the source itself are major sources of error in measurements. First let us consider absorption. For beta and gamma rays this correction can be calculated rather easily, since the decrease in radiation can be taken as exponential. Let us consider the case of a source of thickness t and surface area s (Fig. 6.2). The number of radiations reaching the surface due to an incremental thickness dx at a distance x from the surface can be written as

$$dI = \frac{1}{2} As\rho \; dx \; e^{-\mu_l x} \tag{6.4}$$

where A is the specific activity of the sample, s is the area of the sample, ρ is the density, and μ_l is the linear absorption coefficient. The factor $\frac{1}{2}$ arises because dI is the number of counts at the surface (only half of them pass through one surface) and we are assuming there is no backscattering. Therefore the surface activity due to the whole sample is given by

$$I = \frac{1}{2} As\rho \int_0^t e^{-\mu_l x} \; dx = (As\rho/2\mu_l)(1 - e^{-\mu_l t}) \tag{6.5}$$

In the limiting cases (1) $t \to \infty$, I approaches a maximum value

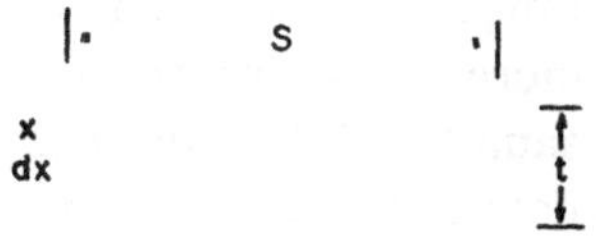

Figure 6-2. A source of thickness t and surface area s.

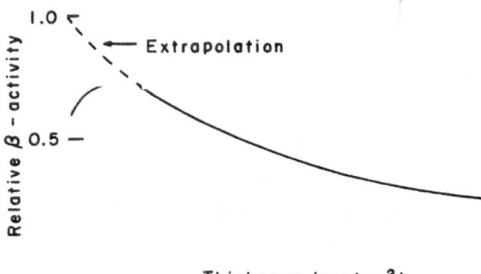

Figure 6-3. The effect of scattering on counting.
The measured activity reaches a maximum for a
source thickness of about 5–10% of the range of
the beta particles.

$As\rho/2\mu_l$ and (2) when t is very small, $I = \frac{1}{2}As\rho t$, which is the activity we should expect when there is no absorption. The relationship between the total activity and the measured activity at the surface can be written in the form

$$I_{\text{measured}} = I_{\text{total}}f_{\text{sa}}$$

where f_{sa} is the correction factor due to self-absorption. From Eq. (6.5) one obtains the correction factor as

$$f_{\text{sa}} = (1 - e^{-\mu_l t})/\mu_l t \qquad (6.6)$$

The effect of scattering in the sample on counting is not simple and it is difficult to derive an equation for it. There have been several attempts[2,3] in this direction and there are also several published experimental curves for the scattering correction factor. However, these are not general enough to apply to all experimental situations. Figure 6.3 illustrates the effect of scattering on the activity measurement. As the thickness increases, the measured activity increases, reaches a maximum for about 5–10% of the range of the beta particles, and then decreases. The initial increase is due to the scattering in the forward direction in the sample. If Eq. (6.5) is used to extrapolate the activity from the data for greater sample thickness, the activity one obtains will be much higher than the actual activity, with a possible error of ~60%. Therefore an experimental determination of the error for each setup, by measuring the activity for different thicknesses, including several measurements for thicknesses below the range of the particle, is recommended. The results of these experiments on counting rate vs. thickness are plotted and this plot can be used to obtain a correction factor f_s for absorption and scattering:

$$f_s = \frac{\text{cpm from thick source}}{\text{cpm from source of zero thickness}}$$

6.1.3. *Correction Factor for Absorption in Window and Air*

Correction should be made for the absorption in the material between the sensitive volume and the source. This material consists of the counter window and the air. If the amount of material is known, the necessary correction term can be estimated. For beta and gamma rays, assuming an exponential decrease in counts with thickness, the window absorption correction term f_w is given by

$$f_w = e^{-\mu_l t} \tag{6.7}$$

where μ_l is the absorption coefficient and t is the thickness of the window, including the equivalent thickness of air. The equation is only approximate since the thickness through which the radiation passes changes with angle. The actual correction term is

$$f_w = \frac{1}{2} \int_0^{-\beta} \sin \beta e^{-\mu_l \sec \beta} \, d\beta \tag{6.8}$$

The expression can be evaluated numerically.

An experimental procedure is usually followed in determining this factor. An absorption experiment is first performed, using Al as the absorber, and the results are plotted on semilog paper. A plot of the number of counts vs. the absorber thickness is not always a straight line, but it is close to a straight line. A tangent is drawn to this curve close to the zero-thickness point (Fig. 6.4).

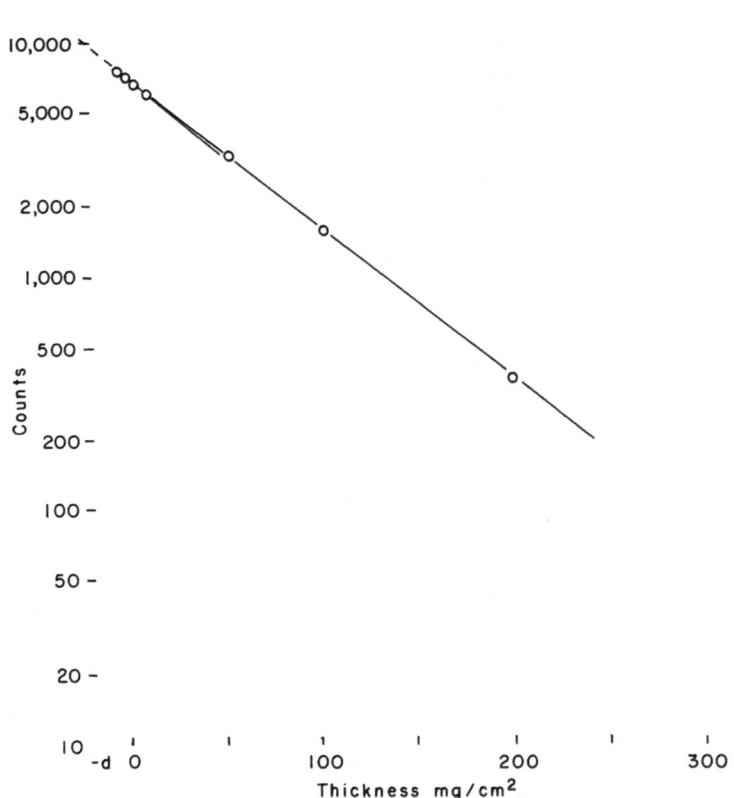

Figure 6-4. Number of counts vs. absorber thickness of ^{210}Bi beta particles. The loss in counts due to window absorption is determined by drawing a tangent to the curve at zero thickness and extrapolating it back to $-d$ (window thickness).

An extrapolation to a negative thickness equal to the average window and air thickness as defined below gives the corrected activity. Since the particles are not all going perpendicular to the window, the path length depends on the angle of incidence. An average thickness can be calculated using the following equation:

$$\bar{t} = \frac{\ln(1/\cos \beta)}{1 - \cos \beta} \, t_w$$

where t_w is the window thickness and β is the angle defined in Fig. 6.1. The difference between $\bar{t}$ and t_w depends on the source-to-counter distance, the difference decreasing as the source-to-counter distance increases.

Because of absorption in air, it is important to have an experimental arrangement of good and accurately reproducible geometry. Let us consider a change dh in vertical positioning. The absorption correction factor changes by an amount $2^{dh/d_{1/2}}$ where $d_{1/2}$ is the half-thickness (2.8 mm in air for ^{14}C radiation). A 1-mm thickness of air corresponds to 0.1293 mg/cm². Therefore the change in counting rate for a 1-mm variation in the positioning of the sample for ^{14}C is $2^{0.129/28}$, which is approximately 4.2%. This error decreases with increase in the energy of the beta rays. For beta rays of ^{32}P ($E_{max} = 1.69$ MeV), the change in counting rate is 0.1% for a change of 1 mm in vertical position. Particular care should also be taken in horizontal positioning of the sample, to avoid errors caused by changes in the geometrical factor.

6.1.4. *Correction Factors for Scattering*

It is important to consider the effect of scattering by air, source cover, source backing, and the detector housing. Correction factors for scattering are always greater than or equal to one, because the scattering tends to increase the number of radiations reaching the detector. The effect of scattering from the source backing on the counting of beta rays has been discussed in Chapter 2.

Now let us discuss the correction factor due to scattering in air. The number of beta particles scattered into the detector per mg/cm² of air depends on the relative position of the source and the detector; the number of scattered particles increases as the layer of air is closer to the source. For this reason it is difficult to estimate the correction factor. One method is to experimentally determine the scattering from thin films of polystyrene at different heights. By measuring the increase in counts for different thicknesses, one first determines the initial slope of the increase in counts. These slopes [increase in counts/(mg/cm²) of polystyrene at zero thickness], determined at different heights, are integrated to give the correction factor. This procedure will give only an approximate value for the correction factor f_a.

The effect of backscattering on the count rate of gamma rays is not as significant as for beta rays. The backscattering factor for gamma rays decreases with energy and with atomic number.

Sometimes the source is covered with a thin plastic or aluminum foil. The amount of radiation absorbed can be estimated from the thickness of the material. The scattering correction is not negligible, since the foil, which is close to the source, scatters a large fraction of the radiation. The scattering correction factor can be estimated as the ratio of counts with the foil (cover) close to the source divided by the counts with the foil close to the window of the counter.

In setting up experiments, one has to be mindful of the effect of scattering from the walls of the source housing. The best procedure is to select the proper materials, preferably of low atomic number, and the dimensions to minimize the effect of wall scattering.

Two additional corrections, the correction for coincidence loss and efficiency, both due to the counting system, have been discussed in previous sections. To make the coincidence correction, one has to know the resolving time of the whole system, which includes the detector and the scaler. Sometimes the resolving time of the system is larger than the resolving time of the detector, especially for the older, slow scalers. From Chapter 4 we can see that the resolving time correction factor f_c is given by

$$f_c = 1/(1 - n'T)$$

Including all the correction factors discussed above, we can write the relationship between the count rate and activity A as follows:

$$\text{count rate} = A \, \epsilon \, G f_c f_s f_w f_b f_a f_h \tag{6.9}$$

where G is the geometrical correction factor; ϵ is the efficiency; f_c is the coincidence correction factor; f_s is the self-absorption and scattering correction factor; f_w is the correction factor for absorption in the window and air; f_b is the correction factor for backscattering; f_a is the correction factor for scattering from air; and f_h is the correction factor for scattering from the housing. All these errors can be estimated from tables or can be determined experimentally. Such errors are called systematic or determinate errors. There is another type of error, called an accidental or indeterminate error. For example, such errors can arise in the measurement of the volume and weight of the sample, during the transfer of the sample, from random noise in the electronic system, and from the erratic behavior of scalers and timers. In radioactive disintegrations there is a built-in source of error in the random nature of the emission of the radiation. Discussion of this is given below.

In the previous sections we have discussed the different sources of error and their effect on the determination of activity. As mentioned earlier, there is no standard set of tables or graphs on errors which can be universally applied. Therefore it is necessary to produce experimentally a set of curves for correction as a function of sample thickness, chemical composition of the sample, sample-to-counter distance, backing material, etc., as required by the experiment. A common type of measurement encountered in analytical laboratories is "relative counting," i.e., the counting of a series of samples containing the same radioactive nucleus but of different activities. If the samples are thin and uniform and are mounted on thin foils, no correction will have to be made in such activities. However, if the sample is thick, corrections must be applied to take into account changes in chemical composition, weight, size, etc. Another approach to this problem is to use sufficiently thick samples, such that the activity reaches the saturation value. In this case reproducible samples can be produced easily.

The above discussion is good for beta particles. In alpha-particle counting the main sources of errors are self-absorption, the absorption in the window, and backscattering. The window (plus air) correction is usually determined experimentally. An equation for the self-absorption correction factor is

$$f_{sa} = \tfrac{1}{2}t/R$$

where t is the thickness and R is the range of the alpha particles in the material. This equation is good for 50% geometry and where $t < R$. By proper choice of the voltage, discriminator setting, and, sometimes, the thickness of the active volume, one can discriminate against low ionizing particles such as beta particles. However, one should be aware of the possibility of pileup of beta pulses, producing pulses of sufficient magnitude to be counted with alpha-particle pulses.

6.2. *ABSOLUTE ACTIVITY DETERMINATION*

A 4π counter (Section 3.5) is a suitable detector for the determination of absolute activity. The only correction that has to be made other than for resolving time is that for the self-absorption. Pate and Yaffe[2] have determined these corrections as a function of thickness and end-point energy. Counting systems with geometry less than 4π (2π or less) can be used for this purpose. However, correction factors become more complex.

Another widely used and simple method is the coincidence method, which is used where two or more radiations are emitted by the nucleus in cascade. Consider a nucleus with a simple decay scheme as shown in Fig. 6.5(a). The nucleus decays to the intermediate state by β-emission. The lifetime of the intermediate state is less than the resolving time of the coincidence circuit. The radiations, β particles and γ rays, are detected by two detectors; one of them is sensitive only to beta particles and the other one is sensitive only to gamma rays. If the activity of the sample is A, the numbers N_β and N_γ of single counts registered by the two detectors are given by

$$N_\beta = A\epsilon_\beta, \qquad N_\gamma = A\epsilon_\gamma$$

where ϵ_β and ϵ_γ are the efficiencies of the two detectors including the geometrical factor. For each beta particle entering the beta counter, the following gamma ray could go in any direction. The chance of the following gamma ray being detected is ϵ_γ and

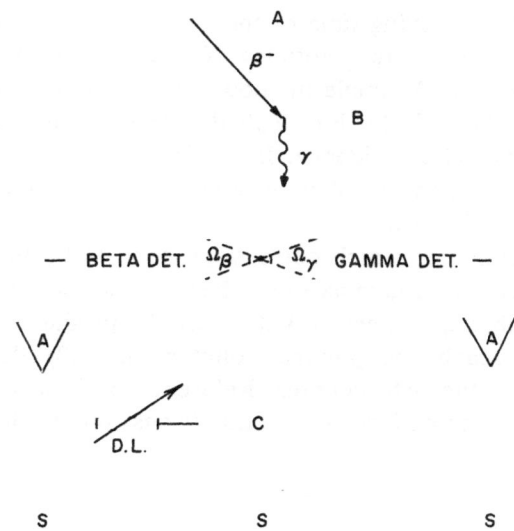

Figure 6-5.   (a) A simple decay scheme showing β-γ decay. (b) Coincidence counting system for absolute activity determination. The sealer S_1 registers the β counts, S_2, the alpha counts and SC the coincidence counts.

therefore the total coincidence count is given by

$$N_c = N_\beta \epsilon_\gamma = A \epsilon_\beta \epsilon_\gamma$$

From the above equations we get for the activity

$$A = N_\beta N_\gamma / N_c$$

Note that the activity is independent of the efficiencies of the detectors. One can obtain the efficiencies from

$$\epsilon_\beta = N_c / N_\gamma, \qquad \epsilon_\gamma = N_c / N_\beta$$

Several corrections have to be made in the above procedure. The counts should be corrected for the background. The coincidence count N_c should be corrected for accidental coincidence. The accidental coincidence rate is given by

$$N_{AC} = 2T N_\beta N_\gamma$$

where $2T$ is the resolving time of the system. If the count rate is very high, it is better to determine N_{AC} experimentally than by using this equation. A simple method is to introduce a time delay in one of the channels (in Fig. 6.5b the delay is shown on the β side) and count the coincidence. If the delay is much greater than the resolving time, the number of counts one gets will be due to accidental coincidences.

Other factors for which corrections need to be made are (1) internal conversion of gamma rays where an electron is produced, (2) Bremsstrahlung, where a beta particle produces an x ray which could reach the gamma counter, and (3) detection of gamma rays by the beta counter. Reference 4 gives an extensive review of this method with a good discussion of the different corrections.

6.3. *RANDOM NATURE OF DECAY PROCESSES*

Nuclear decay processes are random in nature. Hence it is instructive to derive the decay law, given in Eq. (3.1), simply from a consideration of the laws of probability. The basic assumptions for this derivation are: (1) The decay constant λ, denoting the probability of decay per unit time, is constant for all nuclei of the same isotope; and (2) the decay constant λ is independent of the age of the particular nucleus.

If τ is an interval of time small in comparison with $1/\lambda$, the probability of decay in a time τ is $\lambda\tau$. Hence the probability of survival in the time τ is $1 - \lambda\tau$. The probability of survival in a time 2τ is $(1 - \lambda\tau)^2$. It follows that the probability of survival in a time $t = n\tau$ is given by $(1 - \lambda\tau)^n$. In the limiting case where $\tau \to 0$, the probability of survival is

$$\lim_{\tau/t \to 0} (1 - \lambda\tau)^{t/\tau} = e^{-\lambda t} \tag{6.10}$$

Thus the average fraction surviving at time t is $e^{-\lambda t}$. This derivation, based on a consideration of the laws of probability, is due to E. Von Schweidler (1905).

The ultimate accuracy obtainable in nuclear particle counting is limited by the random nature of the nuclear processes. Therefore it is important to be able to compute the uncertainty in a counting experiment. We now review the usual statistical frequency distributions.

6.4. *FREQUENCY DISTRIBUTIONS*

The most fundamental distribution is the binomial distribution, from which the normal, Poisson, and interval distributions can be derived.

6.4.1. *The Binomial Distribution*

If p is the probability for the occurrence of an event and $q = 1 - p$ is the probability for its nonoccurrence, then in n independent trials the probability $P(x)$ that the event will occur x times is given by the p^x term in the binomial expansion of $(p + q)^n$. Hence, since $(p + q)^n = 1$, the expansion represents the sum of the individual probabilities of observing $x = n$ events, $x = n - 1$ events, . . ., $x = 0$ events, that is,

$$(p + q)^n = p^n + np^{n-1}q + \frac{n(n - 1)}{2!} p^{n-2}q^2 + \cdots + q^n$$

Since $q = 1 - p$, any individual term in the expansion can be written in the form

$$P(x) = \frac{n!}{x! \, (n - x)!} p^x(1 - p)^{n-x} \tag{6.11}$$

6.4.2. *The Normal Distribution*

This distribution follows from the binomial distribution in the limiting case where $p = q = \frac{1}{2}$ and $n \to \infty$. In the normal distribu-

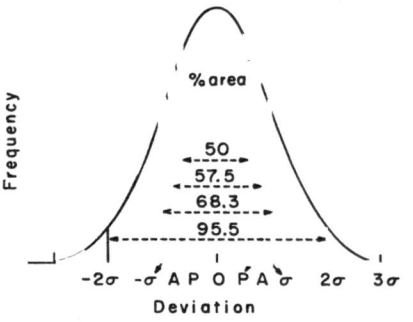

Figure 6-6. Normal distribution function.

tion, the discrete or discontinuous variable x becomes a continuous variable between $-\infty$ and ∞. The normal distribution has two characteristics parameters, the mean value m and the standard deviation σ. In terms of these parameters the probability that x will lie between x and $x + dx$ is given by

$$dp(x) = \frac{1}{\sigma 2\pi} \exp - \frac{(x - m)^2}{2\sigma^2} dx = P(x) \, dx$$

where $P(x)$ is the probability density or probability per unit distance.

A plot of $P(x)$ versus x gives a bell-shaped curve, with the mean value as the most probable value. Figure 6.6 shows some of the main features of the normal curve, in particular, the mean value m and the common measures of precision, namely the standard deviation σ, the average error AE, and the probable error PE. Let x_i represent the value of the ith measurement of a total of n measurements with mean value m. Then these precision measures are defined as follows:

$$\sigma = \sum_{i=1}^{n} (x_i - m)^2 \,/n - 1 \tag{6.12}$$

$$AE = \sum_{i=1}^{n} (x_i - m) \,/n = 0.7979\sigma \tag{6.13}$$

$$PE = 0.6745\sigma \tag{6.14}$$

Often the mean of a set of measurements is represented as $m \pm \sigma$ or $m \pm$ PE. The first indicates that there is a probability of 68.3% that a single measured value in the given set with mean m will not show a deviation from the mean by more than σ. On the other hand, the probable error PE indicates the probability that a deviation of greater or less than PE for a single measured value is 50%. Figure 6.6 shows the numbers 50, 57.5, 68.3, and 95.5. Here the last number indicates that the chance that the measurements will not deviate from the mean by more than 2σ is 95.5%.

6.4.3. *The Poisson Distribution*

This distribution, of major significance in nuclear physics, follows from the binomial distribution for those random processes in which the probability p is very small, that is, $p \ll 1$. Here it is further assumed that the number of events n is large and that the mean value $m = np$ is constant. If m and x are assumed very much smaller than n, it follows that

$$n!/(n - x)! \simeq n^x$$

$$(1 - p)^{n-x} \simeq 1 - (n - x)p \simeq e^{-p(n-x)} \simeq e^{-pn} \qquad (6.15)$$

Hence Eq. (6.11) reduces to

$$P(x) = (m^x e^{-m})/x! \qquad (6.16)$$

For the Poisson distribution the standard deviation for individual observations is $\sqrt{m}$. Hence σ has a definite value in terms of m. The Poisson equation is highly suitable for the case of radioactive disintegration. It can be rewritten in terms of the decay constant λ and the number N_0 of nuclei initially present. In this case $m = \lambda N_0 t$, and the probability of obtaining a count of x in the time interval t is

$$P(x) = [(N_0 \lambda t)^x e^{-N_0 \lambda t}]/x! \qquad (6.17)$$

An interesting application of the Poisson distribution is to determine the efficiency of a Geiger counter. Assume an incident high-energy radiation having a specific ionization, $dE/dx = -270$ eV/cm, a counter gas of argon, and a path length in the counter of 0.5 cm. Since the average energy for producing an ion pair in argon is 27 eV, the average number of ions produced in the counter by the radiation is given by

$$m = \frac{dE}{dx}\frac{d}{w} = \frac{270}{27}\frac{1}{2} = 5 \text{ ion pairs}$$

The probability of producing no ion pair is

$$P(0) = \frac{5^0 e^{-5}}{0!} = 0.0067$$

Therefore the probability that at least one pair will be produced in the counter, which is the efficiency of the counter, is given by

$$P(x > 0) = 1 - P(0) = 1 - 0.0067 = 0.9933 = 99.33\%$$

6.4.4. *The Interval Distribution*

This distribution, derivable from the Poisson distribution, indicates the distribution of intervals of time between successive events occurring in a random process and at a constant rate of events per unit time. The probability that no event will occur in a time interval t, where an average number at is expected, can be obtained from Eq. (6.16) as

$$P(0) = \frac{(at)^0 e^{-at}}{0!} = e^{-at} \tag{6.18}$$

The probability that there will be an event in the interval t to $t + dt$ is $a\,dt$. The joint probability that no event is in the interval t but one event is between t and $t + dt$ is $ae^{-at}\,dt$. Hence the probability that a particular time interval lies between t and $t + dt$, designated by $dp(t)$, is given by

$$dp(t) = ae^{-at}\,dt$$

For a larger number n_0 of intervals, the number of intervals n

greater than the time t_1 but less than the time t_2 is

$$n = n_0 \int_{t_1}^{t_2} ae^{-at} \, dt = n_0(e^{-at_1} - e^{-at_2})$$

In the limiting case $t_2 \to \infty$, the number of intervals greater than the interval t is $n_0 e^{-at}$. Since the average interval is $1/a$, the number of intervals greater than the average interval is $n_0 e^{-1} = 0.37 n_0$. However, if $t_1 \to 0$, the number of intervals shorter than the interval t is $n_0(1 - e^{-at})$.

The above discussion can be used to estimate the coincidence loss in paralyzable counters.* Only if the time intervals between the incident radiations are larger than T, the resolving time (see Fig. 3.12), will the radiations be recorded. The recorded counting rate is therefore given by

$$n' = ne^{-nT} \tag{6.19}$$

where n is the average number of radiations falling on the detector per unit time and n' is the counts registered per unit time. One can also see that the maximum recorded count is given by

$$n'_{max} = 1/eT = 0.368/T = n/e$$

The maximum number of counts is registered when $nT = 1$. A plot of Eq. (6.19) is shown in Fig. 3.13. Coincidence loss in nonparalyzable counters (a good example is a Geiger counter), where events do not affect the counter during its recovery, is discussed in Chapter 5.

6.5. STATISTICAL ERRORS IN NUCLEAR PARTICLE COUNTING

As indicated above, for the Poisson distribution, $\sigma = \sqrt{m}$ and $m = \lambda N_0 t$. It follows that $\sigma = (\lambda N_0 t)^{\frac{1}{2}}$. In actual practice, an

* As the number of incident radiations increases, paralyzable counters fail to produce signals, and consequently the number of counts becomes zero. On the other hand, for nonparalyzable counters, such as Geiger counters, the number of counts increases to a maximum (1/resolving time) as the number of incident radiations increases.

experimenter takes counts for an interval of reasonable duration and the total number of counts N is not expected to correspond exactly to the mean $(\lambda N_0 t)$ for the same duration. Hence m is approximated by putting $m = N$ and in this case

$$\sigma = \sqrt{N} \qquad (6.20)$$

Nuclear counting experiments are always complicated by the presence of background radiation. In practice one counts the combination of source and background N_1 for a time t_1, and then counts the background N_2 for a time t_2. The standard deviation in the total number of counts is $\sqrt{N_1}$ and in the number of background counts is $\sqrt{N_2}$. The counting rate due to the sample, after subtracting the number of background counts from the total number of counts, is given by

$$\text{counting rate} = \frac{n_1}{t_1} - \frac{n_2}{t_2} \pm \left[\left(\frac{n_1^{\frac{1}{2}}}{t_1}\right)^2 + \left(\frac{n_2^{\frac{1}{2}}}{t_2}\right)^2 \right]^{\frac{1}{2}} \qquad (6.21)$$

Note that, statistically, the combined error is the square root of the squares of the individual errors. Let us now consider a specific example where $n_1 = 10,000$ for 10 min and $n_2 = 400$ for 10 min. The counting rate due to the sample is $1000 - 40 \pm [(100/10)^2 + (20/10)^2]^{\frac{1}{2}} = 960 \pm 10$. In the above example the count rate due to the sample is much higher than the count rate due to the background, which is very small. It is evident from Eq. (6.21) that the total error depends on n_1, n_2, t_1, and t_2. If one has a certain total time T for the experiment, what is the proper division of this time between t_1 and t_2 such that the error is a minimum? The time distribution depends on the two counting rates. One can easily obtain an equation for the time for background counts for minimum error:

$$t = \frac{k^{\frac{1}{2}} - 1}{k - 1} T \qquad (6.22)$$

where T is the total, and therefore the time for the combined sample and background counting is $T - t$, and k is the ratio of counting rates $[k = (n_1/t_1)/(n_2/t_2)]$. Let us take the case where $k = 16$, i.e., the sample plus background counting rate is 16 times higher than the background counting rate alone. In this case, $t =$ $[(16^{\frac{1}{2}} - 1)/(16 - 1)]T = \frac{1}{5}T$, or spending one-fifth of the total time in background counting and four-fifths of the total time in the sample counting will minimize the error.

REFERENCES

1. L. R. Zumwalt, Report AECU-567 U. S. AEC, Division of Technical Information.
2. B. D. Pate and L. Yaffe, *Can. J. Chem.* **34,** 265 (1956).
3. B. P. Rayhurst and R. J. Prestwood, *Nucleonics*, **17,** 82 (1959).
4. Comite Consultative pour les Etalons de Measure des Radiations Ionisantes, 5^e session, Gauthier-Villars, Paris (1964).

BIBLIOGRAPHY

E. P. Steinberg, Counting methods for assay of radioactive samples, in *Nuclear Instrumentation and Their Uses* (Arthur H. Snell, ed.), Wiley, New York, 1962, Chapter 5.

Appendix: Selected Constants and Conversion Factors

Speed of light in vacuum, c	2.997925×10^8 m/s
Elementary charge, e	1.60210×10^{-19} C
Avagadro constant, N_A	6.02252×10^{23} mole^{-1}
Mass unit, u	1.66043×10^{-27} kg
	931.478 MeV
Electron rest mass, m_e	9.10908×10^{-31} kg
	5.48597×10^{-4} u
	511.006 keV
Proton rest mass, m_p	1.67252×10^{-27} kg
	1.007276 u
	931.478 MeV
Neutron rest mass, m_n	1.67482×10^{-27} kg
	1.008665 u
	939.55 MeV
Planck's constant, h	6.62559×10^{-34} J-s
Electron volt, eV	1.60210×10^{-19} J
Curie, Ci	3.7×10^{10} dps

Index